A DAVID & CHARLES BOOK

David & Charles is an F+W Publications Inc. company
4700 East Galbraith Road
Cincinnati, OH 45236

First published in the UK in 2007

Conceived, designed and produced by
Quid Publishing
Level 4 Sheridan House
114 Western Road
Hove BN3 1DD
England
www.quidpublishing.com
Designed and illustrated by Lindsey Johns

A catalogue record for this book is available from the British Library.

ISBN-13: 978-0-7153-2696-1 hardback
ISBN-10: 0-7153-2696-1 hardback

Printed in China
for David & Charles
Brunel House Newton Abbot Devon

Visit our website at www.davidandcharles.co.uk

David & Charles books are available from all good bookshops; alternatively you can contact our Orderline on 0870 9908222 or write to us at FREEPOST EX2 110, D&C Direct, Newton Abbot, TQ12 4ZZ (no stamp required UK only); US customers call 800-289-0963 and Canadian customers call 800-840-5220.

a **measure** of all things

THE STORY OF MEASUREMENT THROUGH THE AGES

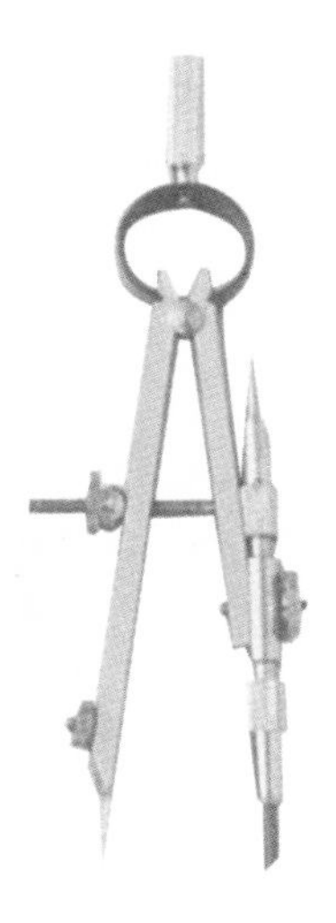

Ian Whitelaw

Contents

'I think that a particle must have a separate reality independent of the measurements. That is, an electron has spin, location and so forth even when it is not being measured.'

Albert Einstein (1879–1955)

'[The universe] is written in the language of mathematics, and its characters are triangles, circles, and other geometrical figures, without which it is humanly impossible to understand a single word of it.'

Galileo Galilei (1564–1642)

Chapter Seven:
A Measure of Time

Chapter Eight:
A Measure of Speed

Chapter Nine:
A Measure of Force and Pressure

Chapter Ten:
A Measure of Energy and Power

Chapter Eleven:
A Measurement Miscellany

Introduction

On hearing that this book was about measurement, a friend perceptively commented, 'Then I guess it's about almost everything!' Methods and units of measurement are certainly relevant to almost every field of human activity.

In the words of Sir William Thomson, Lord Kelvin,

'. . . when you can measure what you are speaking about, and express it in numbers, you know something about it; but when you cannot measure it, when you cannot express it in numbers, your knowledge is of a meagre and unsatisfactory kind; it may be the beginning of knowledge, but you have scarcely in your thoughts advanced to the state of Science, whatever the matter may be.'

What This Book Is

A Measure of All Things is a book about the means that humankind employs, and has employed since the earliest civilizations, to impose order on the world. It is about the systems of measurement that enable us not only to trade in commonly understood amounts and to construct buildings according to agreed standards, but also to investigate, understand, and control the chemical, physical, and biological processes of the natural world, to develop explanatory hypotheses and predictive theories. From art to architecture, and from supermarkets to superconductors, units of measurement are an integral part of our lives, and one that opens a fascinating window on the world.

Intellectual Evolution

To journey through the history of measurement up to the present day is to watch the evolution of human understanding. Weights and measures appear to have been among the earliest tools. Primitive societies needed rudimentary measures for many tasks – constructing dwellings of an appropriate size and shape, fashioning clothing, bartering food or raw materials – and people understandably turned first to parts of the body and elements in their natural surroundings for measuring instruments.

As sedentary civilizations evolved, numbering systems and the science of mathematics developed, and measurement became more complex. The need to understand time and the

m 1 2 3 4 5 6 7 8 9 10 11 12 13 1

1 2 3 4 5

seasons, too, became an important factor for agricultural societies. Whole systems of measurement were created for trade and commerce, land division, taxation, astronomy, and many other branches of scientific enquiry.

Common Understanding

For these more sophisticated uses it was necessary not only to weigh and measure more complex things – it was also necessary to do so accurately time after time and in different places. However, with limited international exchange of goods and communication of ideas, it is not surprising that different systems for the same purpose developed and became established in different parts of the world – even in different parts of a single continent.

In the last few hundred years, steps have been taken to rationalize all units of measurement and to promote a globally shared system. Moving on from the proportions of the human body, barley grains, and the movements of the heavens, the foundations of the international units are firmly laid upon universal constants. Some units of measurement, paradoxically, have their origins in the very nature of energy and matter, in the actual phenomena that we seek to measure, and some even define the limits of what we can measure.

Milestone of Science

The history of measurement is an integral aspect of the history of science itself, and throughout this book we see time and again that the people who introduced new units or discovered relationships between them were also responsible for major leaps forward in scientific understanding. In many cases, units have since been named after these people, and deservedly so.

What This Book Is Not

A Measure of All Things is not intended to be used as a reference book for conversions between units or systems of measurement. Equivalences between various units are given throughout the book, but these should not be relied upon as technical data. The equivalences between certain units are exact by definition; others can be calculated with an accuracy of many decimal places; some are only approximate. We have not made an explicit distinction between these levels of accuracy.

15 16 17 18 19 20 21 22 23 24 25 26 27 28

6 7 8 9 10 11

CHAPTER ONE

systems of **measurement**

Many different systems of measurement have come into being since the beginning of recorded history. Some of the earliest units of measurement drew their inspiration from the natural world, and as a result there are some remarkable similarities between systems that arose independently around the globe. Most modern units strive for natural foundations, taking phenomena such as atomic vibrations and the speed of light as their basis.

'Metric is definitely communist.
One monetary system, one language, one weight and measurement system, one world – all communist!
We know the West was won by the inch, foot, yard, and mile.'
Dean Krakel, Director of the National Cowboy Hall of Fame

Early Forms of Measurement

There is evidence of systems and units of measurement from the earliest civilised cultures, dating back over 5,000 years, suggesting an essential link between social organisation and a practical understanding of the world.

The term *Mesopotamia* (meaning 'between the rivers') was used by the Ancient Greeks to refer to the geographical area between the rivers Tigris and Euphrates, in what is now Iraq. Here, successive cultures developed agriculture, irrigation, and cities over thousands of years – earning the region a reputation as the 'Cradle of Civilisation'. It is also the cradle of measurement, and many elements of the systems developed here to measure time, weight, length, and volume form the basis of units that are still in use today.

Sexagesimal Systems

The first settled dwellers in the fertile lands between the rivers became known as the Sumerians. This is an area that, even at that time, had relatively little rainfall, but the plentiful rivers provided the means to develop wealthy city-states founded on agriculture and trade. Both of these required communication, standardised measurements, and an accurate understanding of the seasons, and the Sumerians' achievements include a system of writing, and mathematics that used several different number systems. It is the Sumerian choice of 10 and 6 as number bases – creating a sexagesimal system (one based on the number 60, which can be subdivided in many ways) – that explains why years, days, hours, and minutes are divided into 12, 24, and 60 throughout the world. In fact, the Sumerian day was divided into 12 hours of daylight and 12 hours of night, although the length of these hours varied throughout the year.

The Measure of Man

Sumerian units of length were largely based on human anatomy, using lengths such as the width of the little finger or thumb, the width of the hand, the distance from the tip of the little finger to the tip of the thumb of an outstretched hand, or the distance from elbow to fingertips. The latter, called a *ku* in Sumerian, and generally referred to as a cubit, is a practical and natural unit. It is seen throughout history in various forms, with differing lengths (from 17 to 25 inches, or 430 to 635 mm) and divided in many different ways (by digits, thumbs, palms, hands, and so on). The equivalents of several Sumerian units were still in use several thousand years later in medieval England.

Sumerian units of area and volume were based on these same units of length, squared or cubed to create two- and three-dimensional units. Larger areas were also based on practical units, such as the amount of land needed to support a family or the area enclosed by an irrigation dyke. Units of weight and monetary value, including the shekel, were also established.

Vitruvian Man:
In this anatomical drawing, Leonardo da Vinci illustrated the human form in the proportions described by Vitruvius, a Roman architect who described the relationships between the width of a finger and the palm, between the palm and the cubit, between the height of a man and the length of his outstretched arms. Da Vinci's drawing is a vision not just of the proportions of man, but of man as a symbol of an ordered and properly proportioned universe.

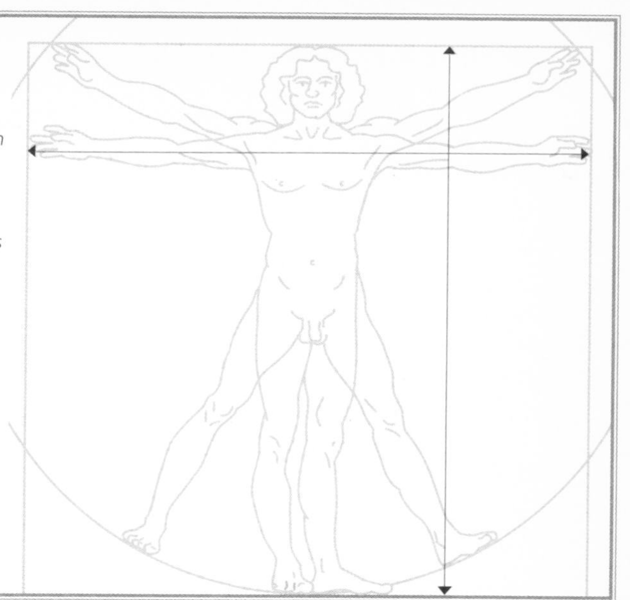

The Need to Standardise

Over time, each of the city-states that made up the region came to define units slightly differently, which posed a problem for a network of trade that was extending over an ever-larger region – the same problem that was to afflict Europe thousands of years later. The solution, in the reign of Gudea (c. 2144–2124 B.C.), was an agreement between city-states on a common standard. Part of this agreement was the acceptance of a unit length that was based on a geographical degree: 1/360 of the circumference of the earth. This was divided into 600 stadia, which were in turn divided into 600 feet. It is probably at about this time that the Sumerian 360-day year came into being.

The Akkadians, who ruled Mesopotamia after the Sumerians, adopted and adapted these systems of weights and measures. The number seven held special significance in Akkadian mythology, and the seven-day week was probably introduced during the second millennium A.D.

On the Banks of the Nile

During this same period, another trading and agricultural civilisation was making its mark to the south, on the fertile banks of the River Nile. Though the Ancient Egyptian system of measurement was borrowed in part from that of Mesopotamia, it largely developed independently and included the use of the number 7 as a base. It, too, introduced units of length based on anatomical proportions, such as fingers, palms, hands, feet, and cubits. Egyptian architects and builders used a set square that had 20/28 of the Sumerian 'Nippur' cubit (called a *remen*) as each of its short sides, and an Egyptian royal cubit as the hypotenuse.

Ancient Mediterranean Measures

In the second millennium B.C., urban centres were established in Anatolia and the Aegean, and by 500 B.C. there was trade between China and Athens.

Given the importance of trade and commerce throughout the Mediterranean region, it is not surprising that a broadly common set of trading standards developed throughout the region, and in the first millennium B.C. the Ancient Greeks used a system of weights and measures that owed much to both the Mesopotamian and Egyptian systems. Later Roman units of measurements were largely based on those of the Greeks, and the Roman system spread throughout much of continental Europe.

Ancient Greek Units of Length

The Greeks used as their basic measure of length the breadth of a finger (*daktylos*), with 4 fingers in a palm (*palaiste*), 12 fingers in a span (*spithame*), 16 fingers in a foot (*pous*), and 24 fingers in a cubit (*pechua*). The daktylos was just over 3/4 inch (19.275 mm), making the Greek cubit a little more than 18 inches or 460 mm long (regional differences in the definition of the daktylos led to some variation in the length of these units, though).

Multiples of these units made up the 6-foot *orguia*, the 600-foot *stadion* (named after the building in which the Olympic running race of this distance was held), and the 5,000-foot *millos*.

Market Measures

By about A.D. 400, the marketplace in Athens had become a centre for trade, and from here the Greek system of weights and measures extended throughout the Mediterranean. The Greeks had separate units of volume for dry goods and liquids (which we find again later in the Imperial and U.S. Customary systems), and standards of measurement in the marketplace were tightly controlled. The Greeks also had a weights system based on the unit of the drachma (1/4–1/6 oz, or 4.5–6 g). The monetary system was based on this same unit in silver.

Ancient Origins

The drachma illustrates how ancient units of measurement can endure, and change form, over the centuries. Having originated as a unit of monetary value in Greece around 1100 B.C., 'drachma' was also found as a unit of weight and of liquid measure – the dram – in England right up to the twentieth century.

Ancient Rome

The Romans were strongly influenced by the Greek system, and adopted many of the same units, although the definitions were not always the same. For example, the *digitus* at the root of the Roman system is slightly smaller than the Greek daktylos, and therefore the Roman foot (*pes*) is slightly smaller than the Greek. The pes is also slightly smaller than the English foot (11.65 inches, or 296 mm). Like the Greek pous, the Roman foot was divided into 16 parts (4 palms of 4 digits each), but by the early Middle Ages in Britain the foot was divided into 12 *unciae*, meaning 'twelfths'. The English 'inch' and 'ounce' are both derived from the Latin word *uncia*, and we'll be meeting the 12-ounce pound a little later (see page 70).

A Roman cubit measured 16 palms (4 Roman feet) and was the distance from a man's hip to the tip of his opposite raised arm. Like the later medieval English *ell*, it offered a convenient way to measure cloth and rope.

The Romans also used units called the *gradus* (step) and the *passus* (pace). These were based on the step of a marching man (the Romans did a lot of marching); the gradus measured 2 1/2 feet, and the passus was two of these. The Roman mile was *mille passuum*, or 'thousand paces', equalling 5,000 feet. Units such as the *pertica* (10 feet) and *actus* (120 feet) were also used.

Other Units

Ancient units of area were derived from the units of length; a square foot was a *pes quadratus*, a square perch was a *scripulum*, and a square actus was an *actus quadratus*. Units of liquid measure ranged from the spoonful and the dose to the *amphora quadratus* (with a volume of 1 cubic foot) and the *culleus* (with a volume of 20 cubic feet). Dry measures were similarly based on the volume of 12 cubic feet – the *quadrantal*, the equivalent of a bushel.

Units of weight rose from the *challus*, which was less than 3/1000 oz (or about 70 mg), through the *scrupulum*, *drachma*, and *uncia* to the *libra*. Most of these units are found later in many parts of Europe, brought to the different cultures by the expansion of the Roman Empire. It is from the Roman libra that the abbreviation lb for pound is derived, for example.

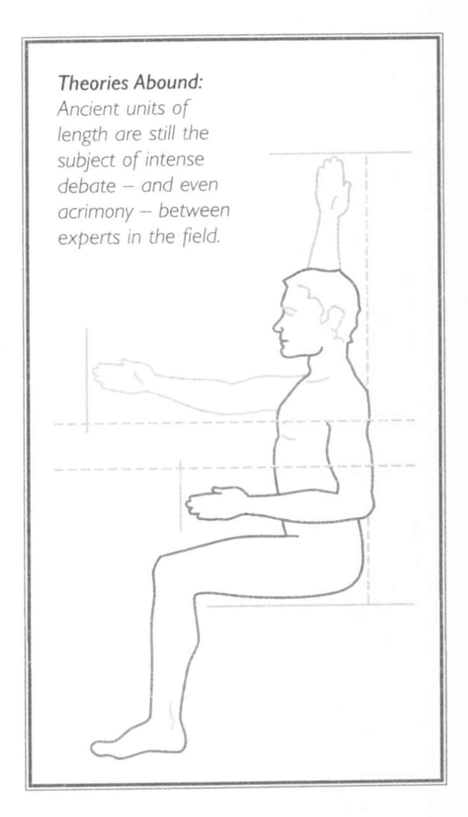

Theories Abound: *Ancient units of length are still the subject of intense debate – and even acrimony – between experts in the field.*

Global Systems

The Imperial and U.S. Customary systems are not the only ones whose line of historical descent seems to stem mainly from Mesopotamia, but this wasn't the only location in which units of measurement came into being.

Several civilisations around the globe independently met the cultural and commercial need for standardised weights and measures with their own – often very similar – systems, based in part on anatomy and the movements of the heavens.

Shared Descent

In the third millennium B.C., to the east of the 'two rivers' in what is now the west and southwest of modern Iran, the Elamite kingdom flourished. Here, too, the systems of the Sumerians and Akkadians were adopted, and their units of measurement – with evolving names and definitions – became those of Ancient Persia. The calendar was also based on that of Mesopotamia. This consisted of a 360-day year divided into twelve 30-day months, with a thirteenth month added every six years to keep the calendar in line with the seasons. In the third century A.D., the 365-day year was introduced, based on the Egyptian solar calendar that had been adopted by the Romans under Julius Caesar.

Persian units of measurement were taken up and developed by the Arabs and became widespread throughout the Islamic world, although the Islamic calendar is in fact a lunar calendar, like that of Judaism.

South Asia

During the third millennium B.C., while the Sumerians were settling in Mesopotamia, yet another major river valley was being civilised – that of the Indus River, in what is now Kashmir and Pakistan. The weights and measures that the 'Harappans' devised and used were both precise and fine, and were some of the first to be standardised. An ivory ruler found in Lothal (one of the main cities of the ancient Indus Valley civilisation, and now one of the region's most important archeological sites) is marked in graduations of about 1/16 inch (less than 2 mm). Mass was measured in units that equated to almost exactly one ounce (28 g), using six-sided weights as small as 1/20 of a unit.

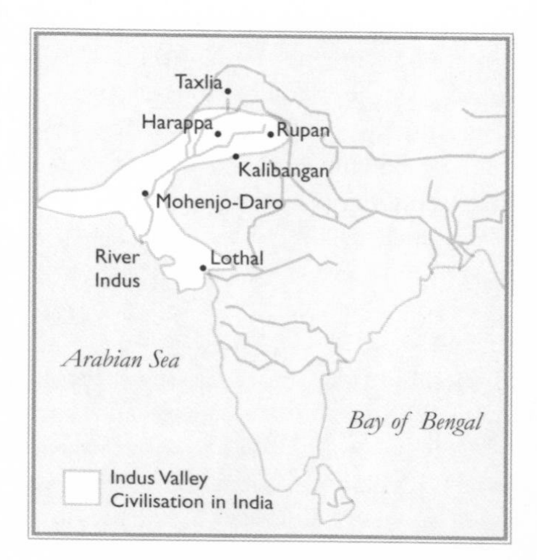

Centimetres

Lothal Ruler: *The Mohenjo-Daro ruler is divided into units corresponding to 1.32 inches (33.5 mm), and these are marked out in decimal subdivisions with amazing accuracy – to within 0.005 of an inch. Ancient bricks found throughout the region have dimensions that correspond to these units.*

East Asia

The Yellow Emperor, who is said to have ruled China in the third millennium B.C., is credited with introducing a consistent system of measurement, with length units based – like those of Mesopotamia – on human anatomy. The inevitable inconsistencies were addressed by King Yu the Great at the end of that millennium, and measuring rules from Shang Dynasty tombs of the thirteenth century B.C. show that at that time a decimal system was already in use for units of length. By the tenth century B.C., the rulers of the Shang Dynasty had put in place a stable and coherent system of units, and this formed the basis for units of measurement later found in Japan, Korea, and South-East Asia.

The Tang Dynasty (A.D. 618–907) presided over a golden age of international trade in China, and the highly defined weights and measures of that period were adopted by Japan at the start of the eighth century. In Japan, this traditional system is known as *Shakkan-ho*, taking its name from the *shaku*, a unit of length equivalent to about a foot (303 mm), and *kan*, a unit of mass of about 8 lb (3.75 kg). Some Shakkan-ho units are still in use, despite the fact that Japan (and China) adopted the metric system in the early twentieth century.

The 'New World'

The ancient and highly advanced cultures of Central America, such as the Maya and the Aztecs, developed complex mathematics and measurements that enabled them to create detailed inventories of property and administer systems of taxation. They also shared a dual calendar in which the year consisted of both a 365-day solar cycle and a 260-day religious cycle that governed divination and ritual observances, including human sacrifice. Evidence of these annual cycles is found in the form of elaborately carved circular calendar stones. The solar and ritual cycles were linked in a 52-solar-year/73-ritual-year cycle (52 x 365 = 73 x 260 = 18,980). Whereas the start of the Ancient Egyptian year was linked to the appearance of Sirius the Dog Star, that of the Aztecs was linked to that of the star group that forms the Pleiades.

Aztec Calendar: *Gods and animals, plants and humans are used to mark the stages of an endless journey through the solar and ritual cycles.*

The English-Speaking World

The direct influence of the Romans' system of measurement during their occupation of Britain seems to have been limited. Certainly Roman coins, and the units of weight to which they were linked, were largely adopted, but this diminished after their departure in A.D. 410.

The centuries after the departure of the Romans saw a steady influx of peoples from many parts of mainland Europe, each bringing with them their own units.

Some of the units the post-Roman invaders brought with them to Britain were indeed descended from the Mediterranean systems, but others had their roots in the agricultural and trading cultures of northwest Europe, with concepts of length and area intimately tied to the land and its use, as we shall see later (see page 28). Between 500 and 1066, these units gradually moved toward becoming a coherent Anglo-Saxon system. In about 960, King Edgar 'The Peacemaker' decreed that all measures throughout the land must agree with standards kept in London and Winchester. From then on, the *bushel*, which was used for grain and agricultural produce, and its subdivisions, became known as 'Winchester measure'. Nonetheless, varying regional definitions continued to abound across the British Isles.

The Arrival of the Normans

In 1066, William of Normandy and his knights effectively took control of England and imposed a truly feudal system on the situation that they found. They brought with them units of measurement that were very largely founded on those of Ancient Rome, but which had many similarities with those of England at the time. Since English measurements were at the heart of land ownership, deeds, records, taxation, and duties, it was not in the Normans' interest to make radical changes to these, and a process of mutual adaptation took place. Units generally kept their Anglo-Saxon values, but were often redefined in Norman terms and given French, Latin-derived names.

Standardisation over the Centuries

With a highly centralised administration operating throughout much of the country, England needed standardised definitions for weights and measures, and successive English monarchs took steps to achieve this.

Legend has it that Henry I (1100–1135) decreed that a yard should be the distance from the tip of his nose and along his arm to the end of his thumb.

The Assize of Measures, issued during the reign of Richard the Lionheart, states that 'Throughout the realm there shall be the same yard of the same size and it should be of iron.'

In the Magna Carta (1215), King John includes a paragraph calling for the standardisation of measurements

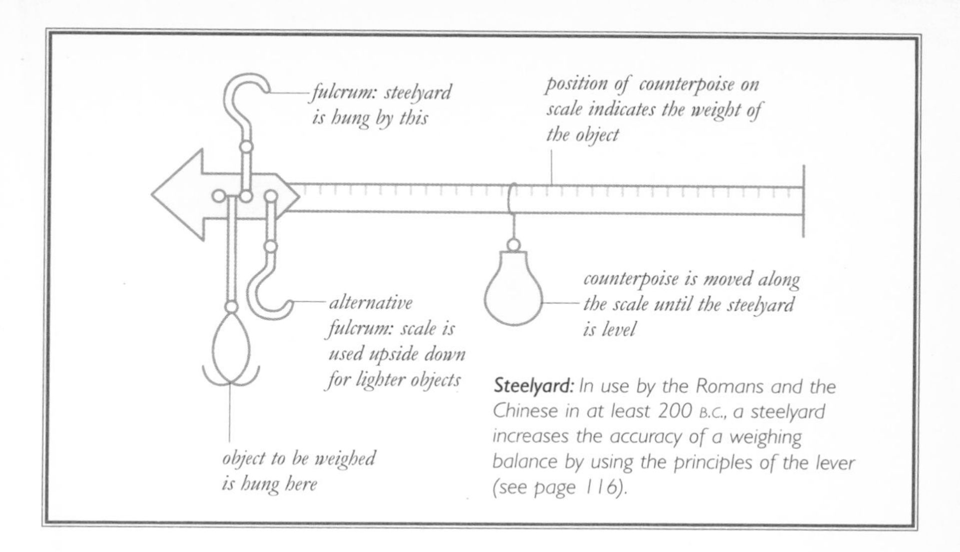

Steelyard: *In use by the Romans and the Chinese in at least 200 B.C., a steelyard increases the accuracy of a weighing balance by using the principles of the lever* *(see page 116).*

throughout the kingdom: 'Let there be one measure of wine throughout our whole realm; and one measure of ale; and one measure of corn, to wit, "the London quarter"; and one width of cloth [. . .] to wit, two ells within the selvedges; of weights also let it be as of measures.'

The subdivisions of the iron yard (or *ulna*, meaning 'arm bone') were defined in some detail in Edward I's reign (1272–1307), as were larger units of length.

In 1357, standardised balances and sets of weights were sent to all the Sheriffs of England, and one of these sets – including a 56 lb (half-hundred-weight), 28 lb (tod), 14 lb (stone) and a pair of 7 lb (cloves) weights – can still be seen in Winchester.

In about 1497, Henry VII ordered an assessment of the country's weights and measures, and dispatched sets of measures across the country, including the bronze bushel, gallon, and quart. Queen Elizabeth I set up a jury in 1574 to examine the standards in use and the lack of uniformity, and she, too, had new standards made and sent out to all the major towns and cities, in 1588. (These remained in use until the introduction of Imperial weights and measures in 1824.) In 1601, new standard measures, including the gallon, quart, and pint, were issued.

In this way, through royal edicts, the system of measurement was adjusted and improved until, by the eighteenth century, the level of standardisation in Britain was probably greater than in any other European country.

This was no accident. During these centuries, Britain had come to dominate international commerce, and had successfully colonised a significant portion of the globe. A coherent system of weights and measures was vital in order to underpin the trade on which British prosperity was founded, and this system was, of course, being spread to all the colonies – those in North America included.

Imperial and Customary

In the late eighteenth century, sweeping changes were taking place in the politics of Britain, France, and the U.S., and the systems of weights and measures in these countries were swept up in this tide of change.

The Imperial System

Fired by an atmosphere of radical change (which was created by the Enlightenment and events such as the French Revolution), scientists in eighteenth-century Europe searched for new ways to base their units of measurement on fundamental phenomena. The consequences were to be far-reaching, but while France was moving toward the metric system, English scientists were looking for accurate ways to determine units in relation to 'natural laws' – the kinds of laws that luminaries such as Sir Isaac Newton had shown were at work in the universe.

Galileo had demonstrated that the period of swing of a pendulum depended on the length of the arm, and that, for example, a pendulum a little more than 39 inches long took one second to make one swing.

The French were to reject this as their definition of length, but in England the relationship between length and period was adopted as a standby definition for the yard. The new definition was based on an engraved brass bar which, if destroyed, would be recreated on the basis of a pendulum swing. The new standard was made legal by the British Parliament in the Weights and Measures Act of 1824. The same Act outlawed several of the old English units, including the troy pound (see page 70), the units of both mass and time were redefined, and volume was defined on the basis of mass. These units formed the basis of the Imperial weights and measures system, also known as the foot–pound–second system.

Several other English units were later discarded as the system was refined and streamlined. The Imperial system became the standard in Britain and throughout the British colonies and the Commonwealth, but not, however, in the Thirteen Colonies of North America, where events had already gained a momentum of their own.

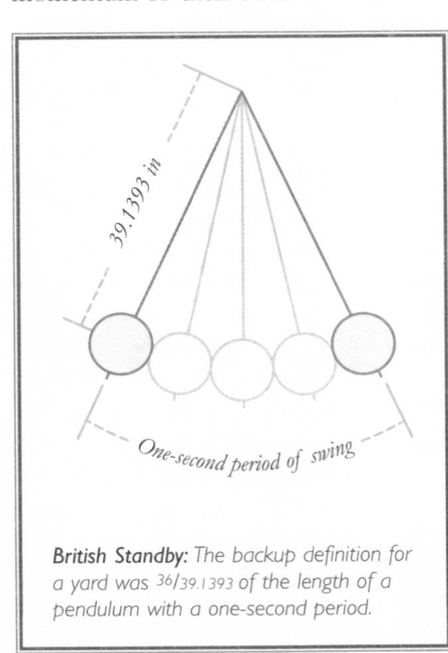

British Standby: *The backup definition for a yard was 36/39.1393 of the length of a pendulum with a one-second period.*

U.S. Customary Units

The American War of Independence achieved its aim on September 3, 1783, with the Treaty of Paris. Under the terms of the treaty, Great Britain formally recognised the United States as an independent nation. At around this time, traditional English units of measurement were in common use, but those very units were seen as emblematic of the yoke that the former colonies had been so anxious to shake off. In the 1760s, Britain had tightened its economic control over the colonies, control that manifested itself in the form of taxation on the basis of weighed and measured goods that were passing through the ports of the eastern seaboard. By the time the British Imperial system came into being, the independent United States had taken standardisation into their own hands.

The need for standardisation to facilitate trade between the colonies had been quickly recognised, and Congress was given the power to rationalise weights and measures in both the Articles of Confederation (1781) and the Constitution of the United States (1790). President Washington referred to the urgency of the matter in his first annual address, Thomas Jefferson recommended a major overhaul, and Secretary of State John Quincy Adams produced a lengthy report that highlighted major shortcomings in the system that was in place. He and Jefferson were both proponents of a metric system, but the changes that did finally come about were simply amendments to the English system. This process lead to the U.S. Customary system, which is still largely in use today.

Thomas Jefferson
1743–1826

John Quincy Adams
1767–1848

Similarities and Differences

The U.S. accepted the new British Imperial yard in the 1800s, and there are few differences between the Imperial and U.S. Customary units of length and area, although an older version of the inch (and hence feet and yards) is used in the U.S. for surveying.

U.S. Customary units of weight are largely the same as the Imperial, with the exception of the hundredweight and the ton, and the U.S. retention of the troy pound, but there are significant differences in some units of volume as a result of the U.S. and Britain choosing alternative versions of the gallon. The U.S. also retained a separate measure for dry volume (see page 58), while the U.K. used the wet-volume gallon for this purpose. Some of these differences have added complications to the U.S. move toward metrication (see page 24). Like Imperial units, the U.S. Customary units of length, mass, and volume are now all defined in terms of the metric standards. The U.S. National Institute of Standards and Technology (NIST) supplies standards to all the States and ensures uniformity throughout the country.

The Origins of the Metric System

As we have seen, systems and units of measurement proliferated throughout history – and nowhere more so than in Europe, where virtually every country had its own system or variation on a system.

By the seventeenth century, the need for a single, coordinated system of measurement was becoming an inescapable reality.

A Revolutionary Step

In 1670, Gabriel Mouton, the Abbé of St. Paul in Lyon, proposed a comprehensive decimal measurement system based on the length of one minute of arc of a great circle of the earth (see page 33). In 1671, Jean Picard, a French astronomer, proposed the length of a pendulum beating seconds as the fundamental unit of length.

However, it was not until 1790, in the midst of the French Revolution, when the National Assembly of France requested the French Academy of Sciences to 'deduce an invariable standard for all the measures and all the weights', that a simple and scientific system was put in place.

The base unit of length was to be a portion of the earth's circumference, and measures for volume and mass were to be derived from this unit of length. The larger and smaller versions of each unit were to be created by multiplying or dividing the basic unit by 10 and its powers. Calculations that could be performed simply by shifting the decimal point were to be a great improvement over the existing systems,

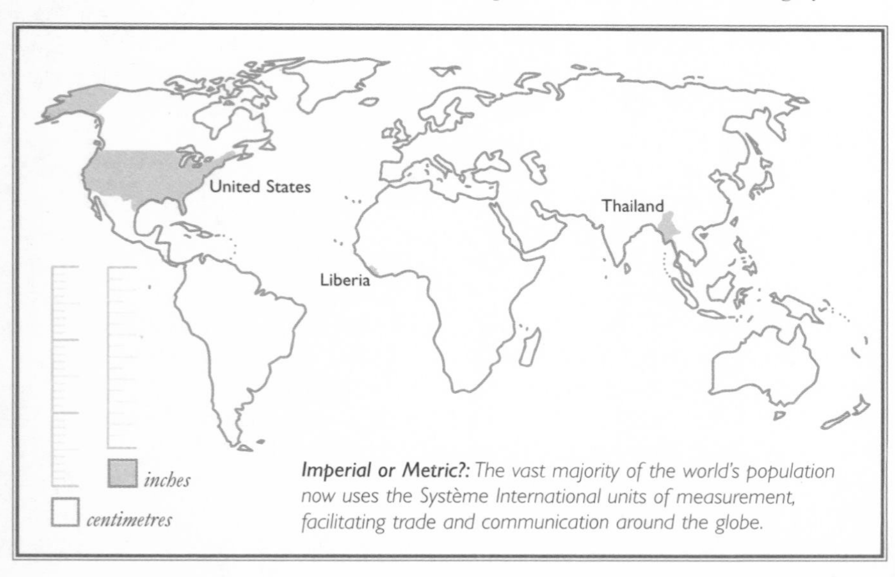

Imperial or Metric?: *The vast majority of the world's population now uses the Système International units of measurement, facilitating trade and communication around the globe.*

A Solid Foundation

In 1960, the General Conference on Weights and Measures revised the system and established seven base units. These are:

the metre	*(m)*	*for length*
the kilogram	*(kg)*	*for mass*
the second	*(s)*	*for time*
the ampere	*(A)*	*for electric current*
the kelvin	*(K)*	*for thermodynamic temperature*
the mole	*(mol)*	*for amount of substance*
the candela	*(cd)*	*for luminous intensity*

Together these base units form the foundation of the Système International d'Unités – the international metric system of units known throughout the world as SI. Today, only a handful of countries – including the United States – has yet to complete the change over to full implementation of SI, and these account for less than seven percent of the world's population. Even in these countries, the metric system is widely used in science, engineering, and medicine.

which required multiplication and division by such numbers as 12, 14, 16, 112, and 1,760.

The unit of length was assigned the name 'metre' (from the Greek word *metron*, meaning 'a measure'), and the physical standard was to be constructed so that it would equal one ten-millionth of the distance from the north pole to the equator along the earth's meridian.

The metric unit of mass – initially called the 'grave' but later changed to the 'gram' – was defined as the mass of one cubic centimetre (a cube that is 1/100 of a meter on each side) of water at its temperature of maximum density. The cubic decimetre (a cube 1/10 of a meter on each side) was chosen as the unit of fluid capacity. This measure was given the name 'litre'.

Gradual Conquest

Despite its many advantages, the metric system was slow to gain international acceptance. However, its standardised character and decimal features made it well suited to scientific and engineering work, and it spread quickly during the nineteenth century – a period that saw rapid technological development.

In 1875, the international Convention of the Meter was signed by 17 countries, and metric standards were constructed and distributed to each nation that ratified the Convention. By 1900, a total of 35 nations had officially accepted the metric system, and by the 1960s, almost all the world's nations had 'gone metric' or were in the process of converting. Notable exceptions include Thailand, Liberia, and the United States.

The Système International

SI is an international system not only because the vast majority of nations have agreed to use it. Adherents to the Meter Convention – first signed in Paris in 1875 – also contribute to its continued improvement through research, meetings, and conferences, and by attending periodic meetings of the General Conference on Weights and Measures to ratify updates to the system.

The Host Nation

As the historical instigator of the metric system from which SI evolved, France is home to the headquarters of the International Bureau of Weights and Measures (Bureau International des Poids et Mesures, or BIPM). As knowledge and technology advance, and ever more accurate measurements become both possible and necessary, the BIPM coordinates the process of refinement of the metric system. Changes to SI – such as the extension of the system to include further units derived from the seven base units, improved definitions, and rules concerning which new or pre-existing units may be used with SI – are approved through the meetings of the General Conference, which meets every four years.

BIPM: *The Convention of the Meter (Convention du Mètre), signed in Paris in 1875, led to the creation of the Bureau International des Poids et Mesures.*

Powers and Prefixes

In the SI metric system, larger and smaller units are derived from the base units by multiplying or dividing by powers of 10. The first power is 10^1 (which is 10); the second power is 10^2 (10 squared, 10 × 10, or 100); the third power is 10^3 (10 cubed, 10 × 10 × 10, or 1,000). When the power is negative it denotes division rather than multiplication, so 10^{-1} means $1/10$; 10^{-2} means $1/100$; 10^{-3} means $1/1{,}000$.

Each power is denoted by a different prefix. For example, multiplying the meter unit successively by 10, the derived units are called the decametre (10 meters), the hectometre (100 meters, or 1 meter × 10^2), the kilometre (1,000 meters, or 1 meter × 10^3), and so on. Dividing the meter successively by 10 to create a scale of ever-smaller units we have the decimetre (0.1 meter, or 1 meter × 10^{-1}), the centimetre (0.01 meter, or 1 meter × 10^{-2}), the millimetre (0.001 meter, or 1 meter × 10^{-3}), and so on. Some prefixes are less frequently used than others – a hectometre is more likely to be expressed as 100 meters, and 10 centimetres is more common than a decimetre.

Abiding by the Rules

As part of its aim of creating and maintaining a system of units of measurement that is both comprehensive and simple, the General Conference excludes the use of non-metric units from the SI system, as they would lead to confusion and inaccuracy. Certain metric units are (sometimes grudgingly) approved

Little and Large

So that the system can be applied to a wide range of phenomena, the list of metric prefixes extends from yotta- at 10^{24} (one septillion) to yocto- at 10^{-24} (one septillionth). The prefixes and their abbreviations, together with the power that they represent and their numerical equivalents, are shown in this table.

Prefix	*Abbreviation*	*Power*	*Numerical equivalent*
yotta-	*(Y-)*	10^{24}	*1 septillion*
zetta-	*(Z-)*	10^{21}	*1 sextillion*
exa-	*(E-)*	10^{18}	*1 quintillion*
peta-	*(P-)*	10^{15}	*1 quadrillion*
tera-	*(T-)*	10^{12}	*1 trillion*
giga-	*(G-)*	10^{9}	*1 billion*
mega-	*(M-)*	10^{6}	*1 million*
kilo-	*(k-)*	10^{3}	*1 thousand*
hecto-	*(h-)*	10^{2}	*1 hundred*
deca-	*(da-)*	10	*ten*
deci-	*(d-)*	10^{-1}	*1 tenth*
centi-	*(c-)*	10^{-2}	*1 hundredth*
milli-	*(m-)*	10^{-3}	*1 thousandth*
micro-	*(µ-)*	10^{-6}	*1 millionth*
nano-	*(n-)*	10^{-9}	*1 billionth*
pico-	*(p-)*	10^{-12}	*1 trillionth*
femto-	*(f-)*	10^{-15}	*1 quadrillionth*
atto-	*(a-)*	10^{-18}	*1 quintillionth*
zepto-	*(z-)*	10^{-21}	*1 sextillionth*
yocto-	*(y-)*	10^{-24}	*1 septillionth*

for use with the system without becoming part of it, but these are gradually being either adopted, with a truly SI definition, or excluded.

The role of SI is to provide the definitions, names, and symbols for the units. Every nation that subscribes to SI must abide by these, but each is free to spell the names of the units in a way that is consistent with its language and culture. To take a simple example, in Denmark, Hungary, and Sweden, the base unit of length is spelled 'meter' – just as it is in the U.S. In Britain, France, Canada, Australia, and New Zealand, however, it is spelled 'metre'. This is not a breach of the SI rules.

The k that we find in km is a symbol for 10^3, not an abbreviation of kilo. Italians are obliged to use the symbol kg to signify kilograms, but they are free to spell the unit 'chilogrammo'.

METRICATION IN THE UNITED STATES

THE SUBJECT OF METRICATION IS NOT A NEW ONE ON THE AGENDA OF THE UNITED STATES. IN 1782, THOMAS JEFFERSON, AS FIRST SECRETARY OF STATE, PROPOSED A DECIMAL-BASED SYSTEM OF WEIGHTS AND MEASURES LARGELY COMPOSED OF DECIMALISED VERSIONS OF THE ENGLISH UNITS.

It was never pursued, but one of the first laws passed under the new Constitution, the 'Mint Act' of April 2, 1792, established the decimal coinage system for the U.S. In the words of the Superintendent of Finance, Robert Morris, '. . . by that means all calculations of interest, exchange, insurance, and the like are rendered much more simple and accurate . . .'

Despite the encouragement of the French and a comprehensive U.S. metric study and report to Congress in 1821, in which John Quincy Adams praised the French metric system, no further steps toward metrication were taken, even though the U.S. had no legal length standard at that time. (A British definition of the yard was adopted by the Treasury Department in 1832, updated in 1855 by a new bronze copy of the Imperial yard sent to the United States by Britain.)

In 1863, representatives of the U.S. government attended two important forums on the issue of weights and measures: the first was the International Statistical Congress, in Berlin, and the second was the Postal Congress, in Paris. The first declared the importance of uniformity in weights and measures, especially for international trade, and the second led to the adoption of the metric system for international mail. In 1866, Congress made it legal to use the metric system throughout the U.S., and defined the yard in relation to a prototype metal bar – the Committee Meter – that conformed to the French Mètre des Archives.

Metric Coinage

U.S. currency, which was already decimal, was assigned metric weights in 1873 (and the current nickel still weighs the same – 5 g).

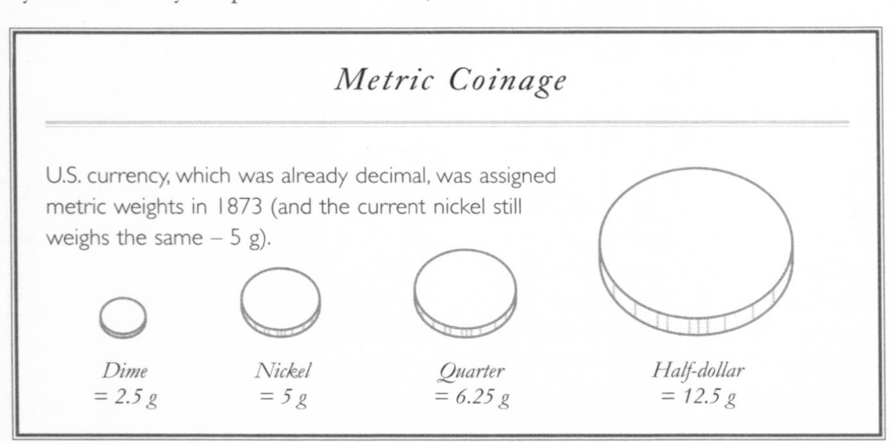

Early Adoption Attempts

In Paris, in 1875, 18 nations signed the Meter Convention, establishing the International Bureau of Weights and Measures, providing for its administration, and effectively taking the first major step toward an international metric system. The U.S. was one of the signatories, and in 1890 it received two copies of each of two metric standards: the International prototype meter and the kilogram. Now held at the National Institute of Standards and Technology (formerly the National Bureau of Standards), one set became the National Prototype Meter and Kilogram, and (under the terms of the 'Mendenhall Order' issued by the Secretary of the Treasury) the primary standards for the United States – not as the basis for U.S. metrication, but as the fundamental standards for determining the yard and the pound. Over the succeeding decades, a series of bills were brought before Congress, all calling for the adoption of metric weights and measures throughout the U.S. – and all were defeated.

A New Impetus

As we have seen, the General Conference on Weights and Measures established the SI system in 1960. It soon became clear to the U.S. that many countries, including some of its main trading partners, were moving inexorably toward acceptance of the new metric system, and in 1968 Congress commissioned a three-year investigation to assess the impact that this would have on the U.S. economy and technological progress. The final report, released in 1971, was unequivocal. It was entitled *A Metric America: A Decision Whose Time Has Come*, and it recommended the implementation of a ten-year plan to make the transition to the predominant use of the metric system.

After a four-year delay, Congress passed the Metric Conversion Act 'to coordinate and plan the increasing use of the metric system in the United States.' The Act made no mention of a time frame, and no legislation was enacted to bring the process about – the conversion was to be voluntary.

In 1976, the U.S. Metric Board was appointed to implement the policy set out in the Act, but, without a mandate to bring about changes or the teeth to enforce them, its efforts were ineffective and largely ignored. The Metric Board was then dissolved in the fall of 1982, raising serious doubts about the U.S. government's long-term commitment to metrication. In the fields of scientific research, medicine, and the military – areas in which international cooperation and communication are central to progress – the SI system was being adopted through necessity, but in the public and private sectors the process was slowing down.

Recognising that metrication is essential in order to maintain competitiveness and conform to international trade standards, Congress gave some encouragement to industry in 1988 by passing the Omnibus Trade and Competitiveness Act. This designated the metric system as the 'preferred system of weights and measures for United States trade and commerce,' and required Federal Government agencies to use the metric system in most of their dealings – thereby hoping to encourage suppliers to do the same.

CHAPTER TWO

a measure of length

Let's start by looking at just one dimension. How long *is* that piece of string? This is a fundamental question. If we can find a way of defining distance in a straight line, there are a host of other definitions that can be based upon this one – sometimes in quite surprising ways.

'He who thinks by the inch and talks by the yard deserves to be kicked by the foot.'
Unknown

Feet, Furlongs, and Rods

Many familiar forms of length measurement took shape in medieval England. If the most ancient units of length are revealed in the monuments and temples of Mesopotamia and Egypt, those of medieval England are inscribed in the very land itself.

The Foot

The foot as a natural unit of length is found in almost every culture, ancient and modern. Initially, it would have been the actual length of a human foot – generally something less than 10 inches (250 mm) – but it was replaced by a longer unit that was more readily divisible by other anatomical lengths, such as the digit, the thumb, the palm, and the hand. The Greek foot was slightly longer than the modern foot; the Roman foot slightly shorter.

In the agricultural society that developed in Britain between the departure of the Romans (410) and the arrival of the Normans (1066), the size of a field was intimately linked to the rights, duties, and taxes that affected the majority of people, and the system of linear measurements reflected this. The natural foot (divided into inches, each composed of three 'barleycorns') and the shaftment were in use, but the key Anglo-Saxon units of length, used to measure out the length and breadth of a field, were the furlong and the rod.

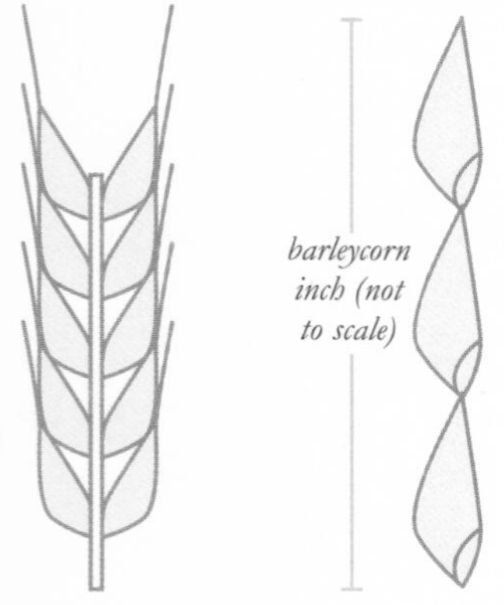

Barleycorn Inch: *While many countries defined the inch as the width of a thumb, in Saxon England it was the length of three barleycorns.*

The Furlong

The word furlong is derived from the Old English *fuhr* (furrow) and *lang* (long), and it was the length of the furrow that a team of oxen could plow without resting. Since this depended to some extent on the nature and consistency of the soil being plowed, the length of the furlong varied across the British Isles. In England, it was the equivalent of 40 rods.

The Rod

The word *rod* comes from the Saxon word *gyrd* or *gyard*, meaning 'straight stick', and the unit's origins probably lie in the length of the plowman's goad used when plowing with a team of oxen. The length of the rod was probably 20 natural feet, and this unit was certainly established by the eighth century.

Manual Foot

In northern Europe at the time of the Roman Empire, the 'manual foot' was a longer version still, corresponding to two 'shaftments'. A shaftment was the length of a fist gripping a pole with the thumb extended, and the manual foot could be measured by holding a stick with both hands, with the tips of the thumbs touching. The shaftment provided a quick way of measuring hand over hand.

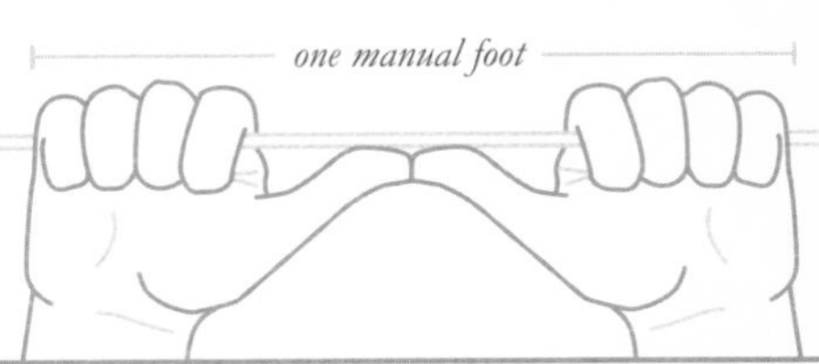

The period after the Norman Conquest saw the reintroduction of some Roman units, but these were integrated with the existing Anglo-Saxon measurements on which landholding and taxes were based. The rod changed its name to the perch (from the French *perche*, Latin *pertica*, meaning 'pole') and, without changing its length, was redefined under Richard the Lionheart as 16½ of the longer Roman feet.

The Iron Ulna

Successive monarchs took further steps to rationalise measurements, and at the end of the thirteenth century, King Edward I had a master yardstick – the 'Iron Ulna' – made as a standard to be used throughout the land. Its length was close to that of the present-day yard, and it was used to define a range of units:

'It is remembered that the Iron Ulna of our Lord the King contains three feet and no more; and the foot must contain twelve inches, measured by the correct measure of this kind of ulna; that is to say, one thirty-sixth part [of] the said ulna makes one inch, neither more nor less . . . It is ordained that three grains of barley, dry and round, make an inch, twelve inches make a foot; three feet make an ulna; five and a half ulna makes a perch (rod); and forty perches in length and four perches in breadth make an acre.'

King Edward I
1239–1307

The unit of three feet gradually became known as the yard (derived from *gyard*, the old word for rod or perch). Given that the perch is defined as five and a half ulnae (yards), this pronouncement defines the furlong (40 perches) as 220 yards.

In North America, canoeists still use the 16½-foot rod as a unit when stating the length of a portage (the distance over which a canoe must be carried between sections of water). The fact that an average canoe is approximately one rod long may account for this.

Miles and Yards

Over the last 500 years, the medieval English units of length evolved into those of both the British Imperial and the U.S. Customary systems of measurement. This development has seen the loss of some units, the addition of a few, and the increasingly accurate defining of most.

Furlong Meets Mile

The Roman mile (the *mille passuum* – 1,000 double steps, making up 5,000 feet) was present throughout the Middle Ages, but its length in England would have been 5,000 natural feet, making it shorter than the original Roman mile. The Roman *mille passuum* was divided into 8 *stadia*, each equal to 625 feet, and it seems likely that the Anglo-Saxon furlong fell into place, in use if not in fact, as one-eighth of a mile.

Edmund Gunter

One-time professor of astronomy at Gresham College, London, Edmund Gunter (1581–1626) was responsible for introducing the terms *cosine* and *cotangent*. Gunter's chain, a mechanical calculating device based on logarithms, was a forerunner of the slide rule. It contained 100 links, with tell-tale marks at 10-link intervals.

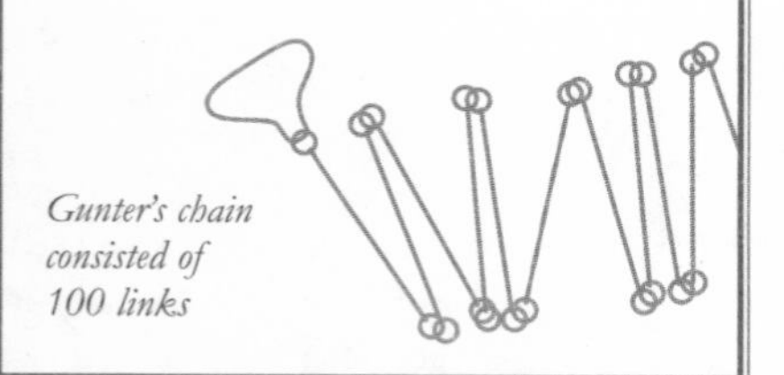

Gunter's chain consisted of 100 links

However, none of the larger units formalised by Edward I would readily divide into 5,000 feet – not the ulna (3 feet), not the perch (16½ feet), and certainly not the furlong (660 feet). This led to two definitions of the mile: one of 5,000 feet, and one of 8 furlongs.

In theory, the furlong could have been redefined to fit. However, because it was fundamental to land ownership, rents, and revenues it could not readily be altered. The solution was provided by Queen Elizabeth I, who replaced the 5,000-foot mile with one of 5,280 feet, or exactly 8 furlongs. A Parliamentary statute of 1592 confirmed this, defining the 'statute' mile as 320 rods (there are 40 rods, or perches, in a furlong) or 1,760 yards.

Rods and Chains

The standard width of a field in Anglo-Saxon England was 4 rods. A rod was 16½ feet, so the width of the field was 66 feet, or one-tenth of a furlong. This distance became a natural unit in its own right, and from the early seventeenth century, a physical chain of this length comprising 100 links was widely used for surveying land. This unit of length was called a chain, and the measuring device became known as Gunter's chain after Edmund Gunter, an English mathematician and surveyor.

The great advantage of the chain was that it worked in a decimal fashion, being divisible by 10 and 100, and 10 square chains equalled an acre (the Anglo-Saxon one-acre field being 40 rods by 4 rods, or 10 chains by 1 chain).

Yard Variations

Although its length did not change significantly, the yard was redefined several times by the creation of successive standard yard bars. The idea of defining it by reference to the length of a pendulum with a certain period of swing was never actually put into action. Definitions of the yard in the U.K. and the U.S. remained in step through successive refinements, with the U.S. accepting the British Imperial yard in 1832 (updated in 1855).

This remained in force in the U.K. until 1958, when the international yard was defined in relation to the metre as being 0.9144 metres exactly (making one international inch measure exactly 25.4 millimetres). This was accepted throughout the Commonwealth of Nations, and in the U.S. Customary system of measurement, but an earlier metric definition of the yard, in which 39.37 inches is exactly 1 metre, was already in force in the U.S. The difference between the two is only about 3 millimetres per mile, but the older definition was so deeply entrenched in surveying and map-making that it has remained in use ever since. For this reason, there are two definitions of the inch, foot, yard, and mile in operation in the U.S.: the Customary System, which is based on the international yard, and the Survey Measure, based on the older definition.

Sports Pitches

Summer games in Anglo-Saxon England were played on fallow fields whose boundaries were marked by ridges of unplowed land, and it was convenient to play across the width of the field. To this day, the length of a cricket pitch (specifically the distance between the two sets of 'stumps') is 22 yards, or 1 chain. 10 chains (220 yards) was the minimum distance over which a long-bow archer was trained to practice.

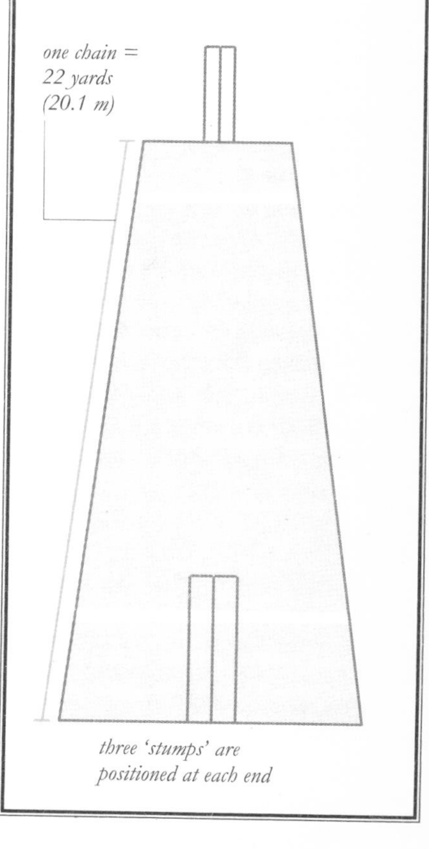

three 'stumps' are positioned at each end

Nautical Units

Seafaring folk have a knack of developing terms that have no application outside the maritime context, and the vocabulary of nautical distance is no exception, although some of the terms began their lives in other fields of activity.

The Fathom

Although best known as a nautical term, the fathom in fact originated in the fields of northwest Europe, where it came to mean a measure of six feet. The word derives from the Old English word *fædm* or *fæthm*, whose meanings include 'embrace', and it is the distance between the fingertips of a man's outstretched arms. Similar words and definitions are historically found throughout Germany, the Netherlands, and Scandinavia. Like the cubit and the foot, this natural, anatomically-based unit of measurement is found under a wide range of names around the globe, from the Japanese *ken* to the Spanish *braza* (from *brazo*, meaning 'arm') and the French *toise*.

Along with the furlong (660 feet), the fathom was used in the measurement of land in medieval Britain, but the rod (into which the fathom resolutely refuses to divide) won out as a division of the furlong (one-fortieth), and the fathom became used exclusively for referring to water depth and the length of ships' cables. A weighted rope with a knot tied in it every fathom was traditionally used to determine, or 'fathom', the depth of water beneath a boat.

After European metrication, two EEC directives allowed the use of the fathom to continue for marine navigation in the U.K. and Ireland, but this authorisation expired at the end of 1999.

Mark Twain

On the Mississippi in the mid-1800s, a depth of two fathoms was called out as 'mark twain' (meaning 'second mark'), which was the minimum safe depth for paddle steamers. The writer and riverboat pilot Samuel Langhorne Clemens (1835–1910) took this as his pen name.

The Nautical Mile

A 'great circle' of the earth is described by the intersection of the surface of the earth with a plane passing through its centre. In other words, if you were to cut the world in half through its middle in any direction, the line of that cut around the world would be a great circle. Great circles that pass through the poles (and intersect the equator) are called 'meridians'. These circles are divided into 360 degrees (degrees of latitude), and each degree is divided into 60 minutes of arc.

Degrees of latitude and minutes of arc are marked on nautical charts, so what could be more sensible than using the minute of arc to define a unit of distance? That is exactly what a nautical mile is. Therefore, if a ship is travelling along a meridian and has changed position by 30 minutes of arc, it has covered 30 nautical miles. And since any straight line over the earth's surface is part of a great circle, it is possible to measure the number of minutes travelled in any direction and determine the distance in nautical miles.

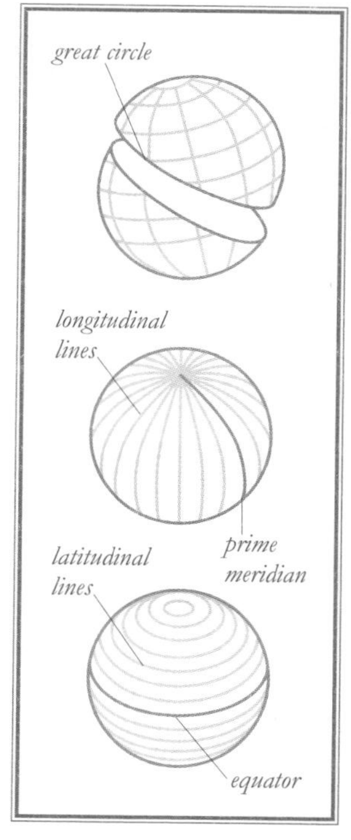

The slight drawback to this system is that the earth is not a perfect sphere, so a minute of arc varies in length depending on your distance from the equator. The original British Admiralty definition was a minute of arc in the English Channel (exactly 6,080 feet, or 1,853.184 metres), but in 1970 this definition was changed to meet the international standard of 1,852 metres (agreed by the International Extraordinary Hydrographic Conference in Monaco more than 40 years earlier).

Although the nautical mile is not an SI unit, it is used around the world for maritime and aviation purposes, as well as in international law and treaties (for example, concerning the limits of territorial waters). There is no standard symbol for the nautical mile, but it is commonly denoted by NM, nm or nmi. The symbol nm also denotes the nanometre in the SI system, but since the nautical mile is 1,852 billion times bigger than a nanometre, there is little chance of confusion.

The Cable

Although very rarely used now, the cable was once a common nautical unit of length. One reason for its demise may have been the confusing multiplicity of its definitions, which included one-tenth of an Admiralty nautical mile (608 feet/185.3184 metres), one-tenth of an international nautical mile (600 feet/185.2 metres), and 100 fathoms (600 feet/182.88 metres), while the U.S. Navy defined a cable as 120 fathoms (720 feet/219.456 metres).

Metre: The Base Unit

The starting point for the entire eighteenth-century French metric system, which was later to develop into the International System of Units (or SI), was a single, fundamental unit of length: the metre.

As discussed in the previous chapter (see page 18), there was an initial attempt to define the metre in terms of the length of a pendulum with a given period of swing, but it was quickly realised that this would vary with location, as the period of swing is dependent on the earth's gravity and the earth is not perfectly spherical.

Instead, it was decided that the metre would be defined as 1/10,000,000 of the length of the earth's meridian from the north pole to the equator (or 1/40,000,000 of the polar circumference of the earth). A seven-year expedition, starting in 1792, proceeded to measure the portion of the meridian between Dunkerque and Barcelona, on which Paris also lies, and produce the exact length of the metre. This was then to be embodied in a prototype metre bar. By definition, the polar circumference of the earth was, at that time, 40,000 kilometres.

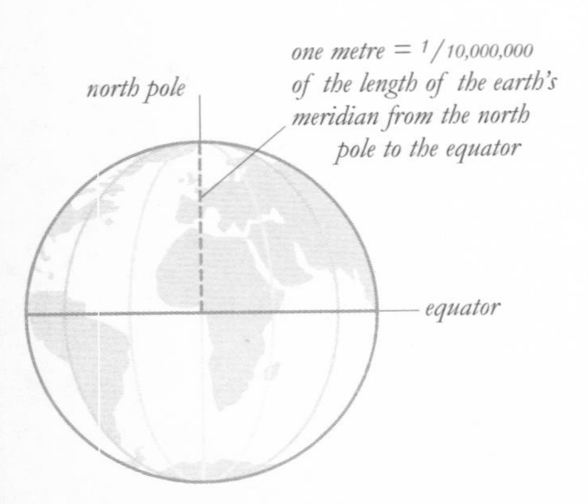

Changing Definitions

The prototype metre bar was found to be incorrect by about one-fifth of a millimetre (the flattening of the earth toward the poles had not been sufficiently taken into account when the original measurements were taken), and this has led to a protracted period of attempts to measure a metre with increasing accuracy.

In 1889, the first General Conference on Weights and Measures established a new International Prototype Metre, as defined by the distance between two lines on a particular bar of platinum-iridium alloy measured at the melting point of ice. This remained the standard until 1960, when the identical length was defined as 1,650,763.73 wavelengths of the orange-red emission line in the electromagnetic spectrum of the krypton-86 atom in a vacuum.

Just in case that wasn't quite precise enough, the metre was redefined by international agreement in 1983 as the length of the path travelled by light in a vacuum during a time interval of 1/299,792,458 of a second.

The table opposite demonstrates how the metric system is derived from the metre, and also shows how the various SI units compare to other systems of measurement. The relationship between different SI units is also shown on page 155.

Units of Length and Their Equivalents

SI Units

Unit	*Divisions*	*U.S. Customary Equivalent*
1 millimetre (mm)		0.03937 in
1 centimetre (cm)	10 mm	0.3937 in
1 metre (m)	100 cm	1.0936 yd
1 kilometre (km)	1000 m	0.6214 mile

SI Units Derived from the Metre

Factor	*Name*	*Symbol*	*Factor*	*Name*	*Symbol*
10^{-1}	decimetre	dm	10^{1}	decametre	dam
10^{-2}	centimetre	cm	10^{2}	hectometre	hm
10^{-3}	millimetre	mm	10^{3}	kilometre	km
10^{-6}	micrometre	µm	10^{6}	megametre	Mm
10^{-9}	nanometre	nm	10^{9}	gigametre	Gm
10^{-12}	picometre	pm	10^{12}	terametre	Tm
10^{-15}	femtometre	fm	10^{15}	petametre	Pm
10^{-18}	attometre	am	10^{18}	exametre	Em
10^{-21}	zeptometre	zm	10^{21}	zettametre	Zm
10^{-24}	yoctometre	ym	10^{24}	yottametre	Ym

U.S. Customary System

International Unit	*Divisions*	*SI Equivalent*
1 inch (in)		2.54 cm
1 foot (ft)	12 in	30.48 cm
1 yard (yd)	3 ft	91.44 cm
1 mile (mi)	5,280 ft	1.609344 km
1 int. nautical mile	2025.4 yd	1.853 km

U.S. Survey Measurement

Unit	*Divisions*	*SI Equivalent*
1 link (unit) (li)	33/50 ft	20.1168 cm
1 foot (survey)	1200/3937 m	30.48006 cm
1 rod (unit) (rd)	25 li	5.029210 m
1 chain (ch)	100 li	20.1168 m
1 mile (survey)	8,000 li	1.609347 km

Measuring the Universe

For measurements on the large scale, we need large-scale units. The distance between stars and galaxies in the universe is so vast that it would be unwieldy to use units such as the kilometre – it would be like measuring the distance from New York to Sydney in millionths of an inch.

Astronomical Units

At a relatively local level – for example, for distances within our own solar system – the 'astronomical unit' can be used. Usually written as AU or au, it is defined as the mean distance between the earth and the sun, which is about 93 million miles (150 million km). This provides a handy way of comparing the orbit of our planet with that of other planets; Mars, for instance, is about 1.5 AU from the sun, while the mean distance from Pluto to the sun is about 45 AU. As soon as we start to look further afield, however, the numbers start to climb. For example, after the sun, the nearest star to us is Proxima Centauri – 268,000 AU away.

Light-Years

When it comes to describing the vast distances often encountered in astronomy, most people are familiar with the 'light-year' (abbreviated as ly). This may sound like a unit of time, but it is actually the distance that light travels in one year.

The speed of light – technically, the speed at which a photon travels through a vacuum – is 670,616,629 mph, or 299,792,458 metres per second (see page 102). As there are 31,536,000 seconds in a year, so a light-year is just under 6 trillion miles in round figures (9,461,000,000,000 kilometres). One light-year is the equivalent of about 63,241 astronomical units, and Proxima Centauri is 4.24 light-years away.

Since the speed of light is known precisely in units of metres per second, the definition of the light-second (the distance travelled by light through a vacuum in one second) is also known precisely. However, since definitions of how many seconds there are in a year vary and there is no single agreed 'reference year', the definition of the light-year is not exact.

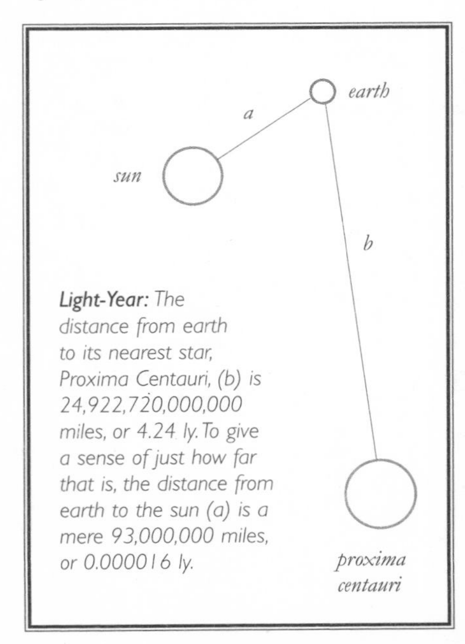

Light-Year: *The distance from earth to its nearest star, Proxima Centauri, (b) is 24,922,720,000,000 miles, or 4.24 ly. To give a sense of just how far that is, the distance from earth to the sun (a) is a mere 93,000,000 miles, or 0.000016 ly.*

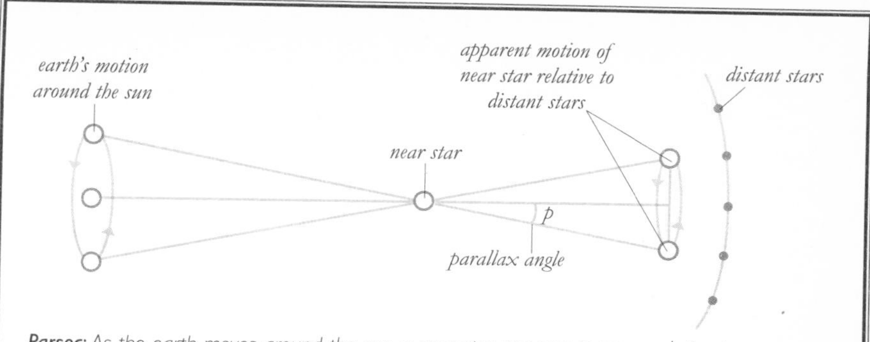

Parsec: *As the earth moves around the sun, a near star appears to move relative to more distant stars. The angle through which it appears to move is the parallax shift. The closer the near star is, the greater this angle will be.*

Parsecs

Most scientists do not, in fact, use light years to define stellar distances. Instead, the preferred unit of measurement for such distances is the parsec, an abbreviation of 'parallax of one arcsecond'. It is defined as the distance at which an object will generate 1 arcsecond (1/60 of an arcminute, or 1/360 of a degree) of parallax (apparent motion relative to another object) when the observer moves one astronomical unit (1 AU) perpendicular to the line of sight. One parsec is equal to approximately 3.26 light-years, but it has the advantage of being directly related to astronomical observations.

To take an example, in the course of 6 months, the earth moves from a position 1 AU on one side of the sun to a position 1 AU on the other side. If in that time a particular star – one to which the line of sight is at right angles to the plane of the earth's orbit – appears to move 2 arcseconds across the heavens in relation to the background stars, then it is at a distance of 1 parsec from the earth. The star has moved 2 arcseconds and the observer has moved 2 AU. The further away a star is, the less parallax is generated, and in fact no star is close enough to generate 1 arcsecond of parallax. Proxima Centauri generates 0.772 arcseconds of parallax and is therefore 1.295 parsecs away.

Beyond the Parsec

One might think that a unit the size of the parsec would be a large enough to deal with the scale of the universe, but that's far from the case. The Milky Way, the galaxy in which our solar system is located, has a diameter of some 30,000 parsecs – and that's still relatively close to home. Fortunately, the metric system of prefixes comes to our aid. 30 kiloparsecs sounds a lot more manageable. However, if we start looking at neighbouring galaxies, we quickly need megaparsecs (1 Mpc = 1 million parsecs, or 3.258 million light-years), and the observable universe is thought to have a diameter of about 8 gigaparsecs (that's 8 billion parsecs or about 26 billion light-years).

Equestrian Measurements

Hands

The English-speaking world – the United States and the United Kingdom in particular – calculates the height of horses in hands. One hand is defined in British law as 101.6 mm (derived from a previous definition of four inches).

The horse is measured from the ground to the top of the withers (the ridge between its shoulder blades), so a horse that is 15 hands high (abbreviated 'hh') is 60 inches from the ground to the top of the withers. Instead of decimal or vulgar fractions, the hand is divided into four one-inch steps, so a horse 62 inches tall is said to be 15.2 hh (pronounced 'fifteen two hands high').

A pony is classified as 14.2 hh or lower.

Like so many measurements, this one clearly originates from the use of the human body – in this case, the width of the palm. That this unit should have endured among horse folk should come as no surprise, given that equestrianism abounds in archaic terms that relate to no other area of life, from the pastern (part of the foot), to the surcingle (a girth that binds a saddle).

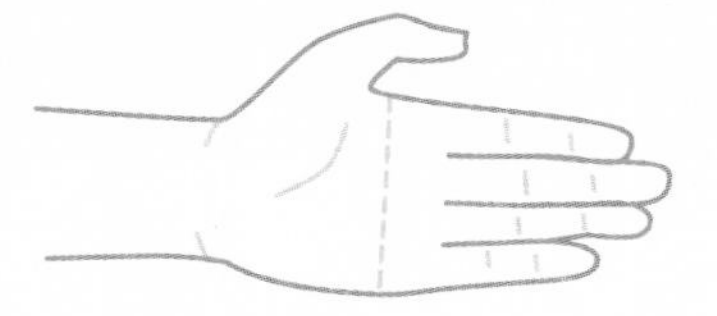

Natural Unit: *The breadth of the hand is still used to measure the height of a horse.*

For Fédération Equestre Internationale (FEI) purposes, a horse can be measured with its shoes on or off. In the United Kingdom, most official measurement of horses is overseen by the Joint Measurement Board (JMB), which requires that the shoes be removed before measurement.

Although the furlong is no longer in common use, distances for thoroughbred horse races in the United Kingdom, Ireland, and the United States are still given in miles and furlongs. Three of the world's most famous horse races are: the Kentucky Derby, which takes place in the U.S. and is run over one and a quarter miles (10 furlongs); the Melbourne Cup in Australia, which is run over two miles (16 furlongs); and, most gruelling of all for those taking part, the Grand National, which takes place in Liverpool in the U.K. and is run over four and a half miles (36 furlongs).

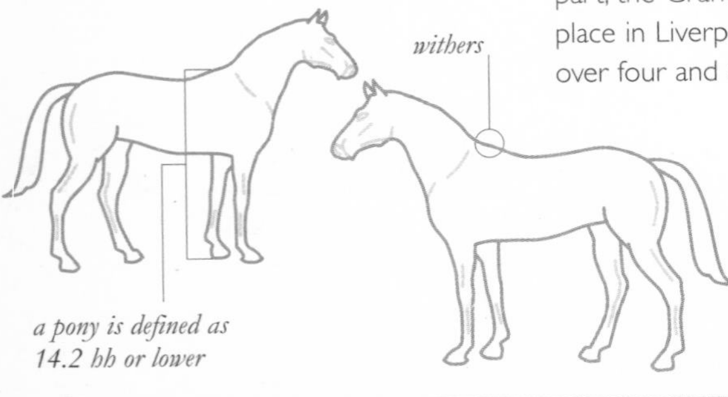

High or Low: *Horses are measured from the ground to the top of the withers (the highest point on the back, on the ridge between the shoulder blades).*

The Limits of Measurement

The first and most fundamental requirement of any unit of measurement is that it can be used to measure accurately and consistently. For this reason there is a limit to how small a unit of length can be. The 'Planck length', developed by and named after German physicist Max Planck (1858–1947), is the smallest unit of measurement with any meaning. At distances smaller than one Planck length, gravity begins to display quantum-mechanical effects, requiring a theory of quantum gravity to predict what happens and effectively rendering the situation unmeasurable.

Planck length is roughly equal to 1.6×10^{-35} metres. Written out, that would be the number 16 preceded by a decimal point and 34 zeros. In relation to subatomic particles, the Planck length is about 10^{-20} times the size of an electron – the kind of particle that orbits around the nucleus of an atom.

'Planck time' is the length of time it would take a photon travelling at the speed of light to cover the distance of a Planck length. Planck time is effectively the smallest measurement of time that has any meaning, and it is equal to 5.391×10^{-44} seconds. In theoretical physics, it is not possible to know what happened less than one Planck time after the Big Bang.

The shortest period that has actually been measured is in the order of attoseconds. An attosecond is 10^{-15} seconds. This is vastly greater than a Planck time (in fact, it is about 1,026 Planck times), but if one attosecond were to last for a second, one second would last for 32 million years.

The pre-metric unit of length on the atomic scale is the ångström (Å), named after Anders Jonas Ångström (1814–1874), a Swedish physicist. One ångström is 10^{-10} m, or 0.1 nanometres. The Bohr radius – the radius of the orbit of an electron at the lowest energy level – is approximately 0.53 ångströms.

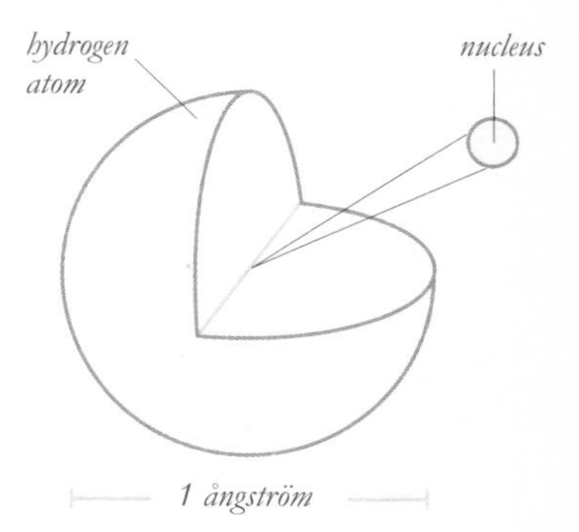

Simplest Atom: *A hydrogen atom, which has just one proton in its nucleus and one electron orbiting at the lowest energy level, has a diameter of just over 1 ångström.*

Planck Length: *The Planck length is roughly equal to 1.6×10^{-35} m. Written out, that is a decimal point followed by 34 zeros and then 16.*

.0000000000000000000000000000000000016

Gauging Thickness

The *Oxford English Dictionary* tells us that *gauge* means: 'A fixed or standard measure or scale of measurement, the measure to which a thing must conform; especially a measure of the capacity or contents of a barrel, the diameter of a bullet or the thickness of sheet iron'.

The key words here are 'a thing', because when it comes to gauges, virtually anything can have its own scale. Almost every trade and craft seems to have invented its own gauge, and many of these are internally consistent.

Sheet Metal

Let's take sheet metal as an example. The thickness of a sheet of metal is defined by a gauge number between 3 and 36. Somewhat confusingly, the thickness of sheet metal decreases as the gauge number increases. A sheet of 10-gauge standard steel is 0.1345 inches (3.4163 mm) thick; if it were 20-gauge, then it would be 0.0359 inches (0.9119 mm) thick. Which all seems straightforward enough. However, do make sure that the sheet isn't galvanised steel or aluminium, because in the first case it will be thicker, and in the second it will be thinner.

Wire Gauge: This tool is used to determine the gauge of an individual strand of wire by finding the slot (not the hole) into which it just fits.

Wire Thickness

The gauge system used to specify the thickness of wire also works in reverse, with a larger number signifying a smaller diameter. One possible explanation for this is that, in the past, wire was made by drawing it through successively smaller holes, and the gauge system may refer to the number of holes through which a wire was drawn.

Very thick wire (between about one-half and one-third of an inch in diameter) has a gauge of 0000000 (or 7/0) down to 0 (or 1/0). The next size down is 1-gauge, and the scale goes down into the 50s.

Unfortunately, the exact thickness of any particular gauge of wire will depend on which of several systems is being employed. The American Wire Gauge (AWG, also known as Brown & Sharpe) and the Imperial Standard Wire Gauge (SWG) are commonly used, but there are others.

To avoid confusion, it is safer to give the thickness of wire in thousandths of an inch, or use the International Metric system, in which the gauge number is the diameter in millimetres multiplied by 10 – a system in which the gauge goes up as thickness increases. 20-gauge wire has a diameter of 2 mm in the metric system . . . unless you're using the AWG Metric System, in which case the wire has a diameter of 0.8128 mm.

Shotgun Barrels

In yet another successful attempt to create a system in which more means less, shotguns and shotgun shells are given a gauge designation that goes down as the diameter, or bore, of the barrel goes up: 20-gauge, 16-gauge, 12-gauge, 10-gauge, and so on. The number in each case refers to the number of lead balls, each with a diameter equal to the internal diameter of the barrel, that it would take to make up a pound of lead. The larger the barrel, the fewer lead balls it takes.

The odd one out in the shotgun system is the '.410', or four-ten – a designation that is not a gauge but a calibre. This means that it refers directly to the barrel bore of 0.410 inches.

Between the Tracks

The distance between the rails on a railroad is also referred to as the gauge. When the first modern railways were being built in Britain in the early 1800s, George Stephenson chose the common width of the coal-wagon rails – 4 feet 8 1/2 inches (1,435 mm) – as his gauge. Isambard Kingdom Brunel, on the other hand, and at the other end of England, chose a wider gauge of 7 feet 1/4 inch (2,140 mm). Although the wider gauge had many advantages – not least of all, stability – when the Government recognised the need to standardise the width there was far more of Stephenson's track in use, and so his narrower gauge was made the standard. This width is now is used in much of the world.

Brunel's gauge = 7 ft 1/4 in (2,140 mm)

Stephenson's gauge = 4 ft 8.5 in (1,435 mm)

Model-railway enthusiasts work in a variety of scales that are also referred to as gauges. An O-gauge model railway has a scale of 1:43.5, or, as it is sometimes (unhelpfully) described, 7 mm to one foot.

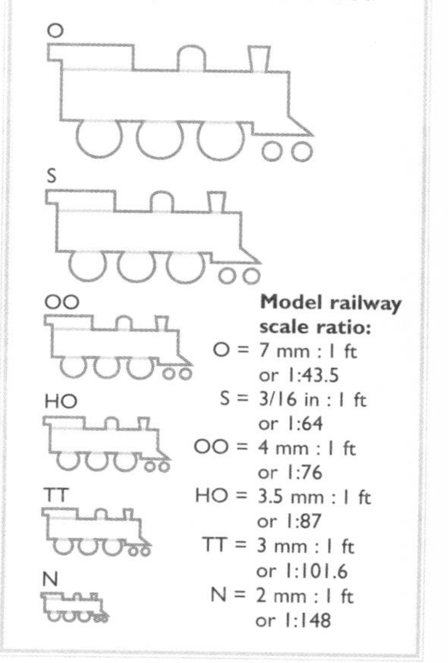

CHAPTER THREE

a measure of **area**

It's all very well measuring in straight lines, in one dimension, but in ancient matters of practicality such as land ownership, agricultural work, and taxation, area is an important concept. Multiply the length by the breadth and you have the area of the item in question – the field, the landholding, the sheet of paper – in square units.

'Now would I give a thousand furlongs of sea for an acre of barren ground.'
William Shakespeare, The Tempest

Medieval Units of Land

Like units of length, units of land area have their origins in practical considerations. Some have ancient roots, while others refer to farming methods and the feudal system.

The Acre

The word 'acre' has a long history, coming to us from the Sanskrit (in which *ajras* was used to describe open or untenanted land), through Greek, Latin (the word *ager*, from which 'agriculture' is derived, means 'field'), Old High German, Saxon, and Old English. To the Saxons, too, the acre was a field, but one of specific size and proportions.

We have already seen (pages 28–9) that the basic units of length in Anglo-Saxon England were the furlong and the rod, and that the size of a field was one furlong (220 yards) by 4 rods (22 yards). This was the definition of an English acre – 4,840 square yards – and it was seen as the area of land that a medieval plowman could work in one day. Since the width of the field was four times the length of the plowman's goad (*gyard* or 'rod'), he could readily determine what proportion of his day's work was done. In fact, the term 'rod' was also used to refer to one-quarter of an acre, or a strip one furlong long and one rod wide.

Just as the furlong differed in length in various parts of Britain (due to differing soil conditions), so did the acre. In Scotland, where the soil was easier to plough, it was over 6,000 square yards, and in Ireland it was almost 8,000.

The Hide

The land in a Saxon village was parcelled out as fairly as possible, and each family would have fields in different parts of the district so that no one had all the best-quality land. The amount of land judged to be enough to support a family was called a 'hide'. In England this was about 120 acres, although the actual size varied widely, the hide being defined by its productivity rather than its area. The hide was an important unit for taxation. It was sometimes regarded as being the same as a 'carucate' – the amount of land that a team of eight oxen could plow in a season. One-quarter of a carucate was called a 'virgate', and one-eighth was a 'bovate' – the area that one ox (or *bos*) could plow in a season.

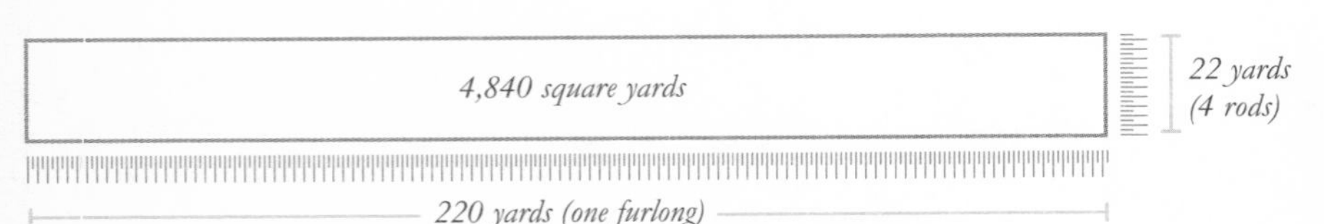

Acre: *The long, thin shapes of typical Saxon one-acre fields can still be seen in many parts of rural England, even though they have long been plowed under. The boundaries are revealed as slight ridges, and the fields themselves make a patchwork of faintly differing colours.*

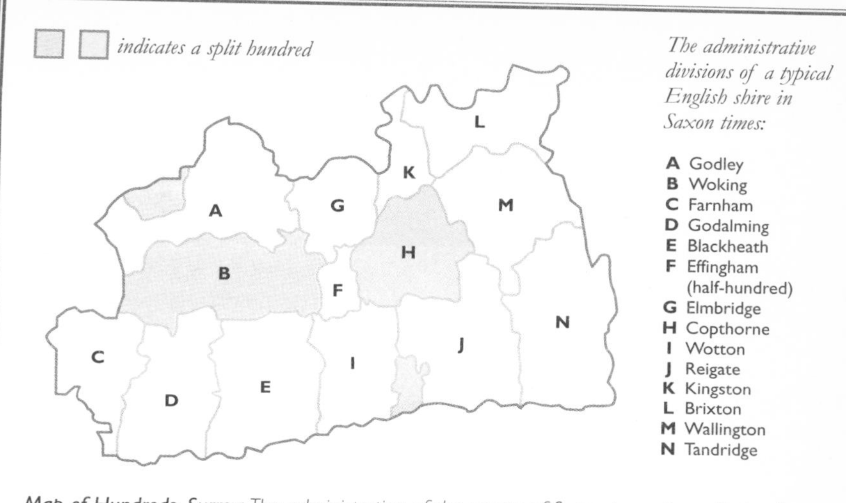

Map of Hundreds, Surrey: *The administration of the county of Surrey, in southern England, was based on geographical divisions known as 'hundreds' from Saxon times until the late nineteenth century. Each hundred was divided into tithings and townships, which retain their identity today as parishes. Each of the hundreds held its own monthly court, called the hundred-moot, and the county held a court called the Shire-moot twice a year, presided over by the Ealdorman and the Shire Reeve (from which the word 'sheriff' is derived).*

Knight's Fees and Hundreds

In the medieval feudal system, the use of land owned by the nobility was granted as a 'fief' (hence the term 'fiefdom' as the estate or domain of a feudal lord) to vassals. They worked the land and made their living from it, and in return were expected to provide military service or the equivalent in monetary value. The service due was in direct proportion to the value of the land, which was therefore calculated in units of the knight's fee, which generally equated to five hides. For every knight's fee of land, the landholder was expected to provide the means to support a fully armed knight and his horse, and keep him in the manner to which he was accustomed. A single fief might consist of a fraction of a knight's fee, or many tens of these units. Larger fiefs were divided into subfiefs, and in practice the payment of the knight's fee would be raised not by one single landholder, but by all the families working the land.

For administrative purposes, hides were grouped in units of one hundred called, logically enough, 'hundreds'. The hundred – divided into ten tithings, each one consisting of ten households – was the unit of which the larger administrative unit of the shire was composed, and the basis on which the community's military service and taxation were assessed. Primarily found in England and Wales, hundreds were also used as an administrative division of land in parts of Scandinavia and the U.S. In the north of Saxon England, hundreds were known as 'wapentakes'.

Units of Area

ALTHOUGH THERE ARE SOME SPECIALIST UNITS OF AREA – SUCH AS THE ACRE – IN THE MAIN, BOTH THE U.S. CUSTOMARY AND BRITISH IMPERIAL SYSTEMS BASE THEIR UNITS OF AREA ON THE SQUARE OF THEIR LINEAR UNITS.

Officially, SI has largely replaced the Imperial system around the world, but common units such as square inches, feet, yards, and miles, are still widely found, especially in the U.S. For example, the size of a room or building is usually described in terms of square feet; flooring material is usually bought by the square yard; and square miles are used to measure territory. The acre is still commonly used in reference to property. In surveying, additional units such as the square rod and the square chain are also used (although rarely) and, due to the continued use of the Survey measure, the size of the survey square mile is fractionally different from the international definition.

The tables below show common area measurements and their equivalents.

International Units of Area:

Unit	*Divisions*	*SI Equivalent*
1 square inch (sq in)		6.4516 cm2
1 square foot (sq ft)	144 sq in	929.0304 cm2
1 square yard (sq yd)	9 sq ft	8,361.273 6 cm2
1 square mile (sq mi)	3,097,600 sq yd	2.58998810336 km2

U.S. Survey Units of Area:

Unit	*Divisions*	*SI Equivalent*
1 square foot (sq ft)		929.0341 cm2
1 square rod (sq rd)	272.25 sq ft	25.29295 m2
1 square chain (sq ch)	16 sq rd	404.6873 m2
1 acre	10 sq ch	4,046.873 m2
1 square mile (sq mi)	640 acres	2.58998811 km2

SI Units of Area:

Unit	*Divisions*	*U.S. Customary Equivalent*
1 square centimetre	100 mm2	0.1550 sq in
1 square mete	10,000 cm2	1.1960 sq yd
1 square kilometre	100 ha (not SI)	0.3861 sq mi
1 hectare (not SI)	10,000 m2	2.4711 acres

Homesteads and Townships

The U.S. Congress passed the Homestead Act in 1862 with the aim of encouraging migration toward the West and settling new territories, and similar legislation was later passed in Canada and Australia. Working from a grid of north–south meridians and east–west baselines, the land was surveyed and parcelled out for settlers, and this patchwork is still visible on the landscape of North America. Indeed, the Homestead Act was only repealed in 1976 (1986 in Alaska).

Working from the base grid, the land was subdivided by parallel lines six miles apart. Those running east–west were called township lines, and those running north–south, parallel to the meridians, were called range lines. These created a grid of squares with sides of six miles, and each 36-square-mile block was called a township.

Each township was divided into 36 sections of one square mile each. These were numbered 1 to 36 from the top right (northeast corner) to the bottom right (southeast corner) in a zigzag sequence. Since there are eight furlongs to a mile, and an acre is one-tenth of a square furlong, each square mile section had an area of 640 acres.

Each section was divided into four quarters each with an area of 160 acres. The Homestead Act gave one-quarter of a section of undeveloped land to anyone over the age of 21, on condition that they lived on it for five years and built a house. Alternatively, they could buy the land for $1.25 per acre after six months.

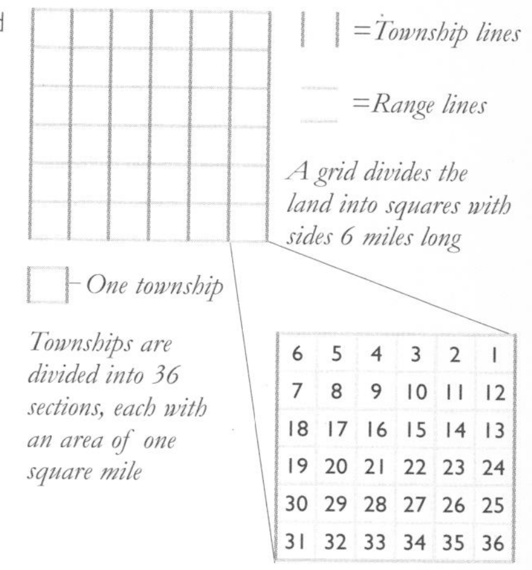

Grid Systems: *In many parts of North America, roads have followed the network of township and range lines, giving the map its familiar grid-like appearance.*

Subdivisions

Further subdivisions of the quarter create quarters of quarters (40 acres each) and quarters of quarters of quarters (10 acres each). The precise location of each subdivision is specified on land deeds, using a system that identifies the township (by the number of the township line and range line, and their direction from the base line and meridian), the section (by its number), the position of the quarter within the section (by its compass position), and the position of any further subdivision (in the same way). This leads to definitions such as: The Southwest Quarter of the Northeast Quarter of the Southeast Quarter of Section 34, Township 11 South, Range 2 West.

Metric Area

Rather than creating entirely new units for the purpose, the measurement of area in the metric system is based on the same units that are used to measure length, starting with the metre as the base unit.

The Base Unit

The area of a rectangle is the length of one side multiplied by the length of the adjacent side. Whatever the linear units used to measure the length of the sides, the area will be expressed in those units squared. As we have seen, the basic SI unit of length is the metre (which is currently defined as the length of the path travelled by light in a vacuum during a time interval of 1/299,792,458 of a second). The derived SI unit of area is, therefore, the square metre or metre squared (m^2). A square that has sides one metre long has an area of one square metre (1 m^2 or, less formally, 1 sq m).

Other Units

The beauty of the SI units and the system of prefixes that allows the basic units to be multiplied by powers of 10 is that the same can be done to all the derived units, such as the square metre. Therefore, a square with sides 1 cm long has an area of 1 cm^2.

It is important to bear in mind that the area of a rectangle is proportional to the square of its linear dimensions. For example, when the lengths of the sides of a rectangle are doubled, the area is quadrupled. For this reason, the power of each of the square units is twice that of the linear unit with the same prefix.

Metric Units of Area:

Multiple	*Name*	*Symbol*	*Multiple*	*Name*	*Symbol*
10^0	square metre	m^2			
10^2	square decametre	dam^2	10^{-2}	square decimetre	dm^2
10^4	square hectometre	hm^2	10^{-4}	square centimetre	cm^2
10^6	square kilometre	km^2	10^{-6}	square millimetre	mm^2
10^{12}	square megametre	Mm^2	10^{-12}	square micrometre	μm^2
10^{18}	square gigametre	Gm^2	10^{-18}	square nanometre	nm^2
10^{24}	square terametre	Tm^2	10^{-24}	square picometre	pm^2
10^{30}	square petametre	Pm^2	10^{-30}	square femtometre	fm^2
10^{36}	square exametre	Em^2	10^{-36}	square attometre	am^2
10^{42}	square zettametre	Zm^2	10^{-42}	square zeptometre	zm^2
10^{48}	square yottametre	Ym^2	10^{-48}	square yoctometre	ym^2

Cosmic Scale: *Given that its sides are about 100 million light-years long, the square yottametre is a little used unit; mm^2 through to nm^2 are more common.*

What Happened to the Hectare?

Prior to the introduction of the Système International in the 1960s, a variety of metric systems of weights and measures were in use throughout continental Europe, and some of these had separate base units for dimensions that are actually geometrically related. In the original French system, for example, the metre was used as a unit of length but the 'are' (rather than the square metre) was used for area. This was, and still is, an area equivalent to 100 square metres, or the area of a square with sides 10 metres long.

As in the SI system, the prefix 'hect-' indicates a power of 100, and a hectare is equal to 100 ares (10,000m^2 or the area of a square with sides 100 metres long). In the SI system, a hectometre is, of course, 100 metres, so a hectare is equivalent to a square hectometre. As a unit, the hectare is completely in line with the SI system in both its metric definition (in terms of an SI base unit) and its prefix, and it is therefore accepted for use with the SI system – but it is not an SI unit because it originates from an older form of the metric system.

The hectare (which is the equivalent of about 2.5 acres) remains a common unit, and is used in the contexts of real estate, agriculture, and town planning. It is also used – in the form of 'millions of hectare metres' – to measure the capacity of water reservoirs. In Germany, France, and Spain the are and the decare (10 ares, 1,000 m^2, or $^1/_{10}$ of a hectare) are also encountered, but the are is rarely used with any other prefixes.

Perhaps the most famous example of a hectare is Trafalgar Square in London. Laid out in commemoration of the Battle of Trafalgar (1805), this is the city's only metric square.

Nelson's Column stands 46 metres above London's only metric square, which measures one hectare.

A decametre is 10 metres, but a square decametre is 100 (i.e. 10 x 10, or 10^2) square metres. A kilometre is 1,000 metres (1 metre x 10^3), but a square kilometre is 1,000,000 (1,000 x 1,000, or 10^6) square metres. The table on the facing page shows the relationship between the square metre and each of the other metric units of area.

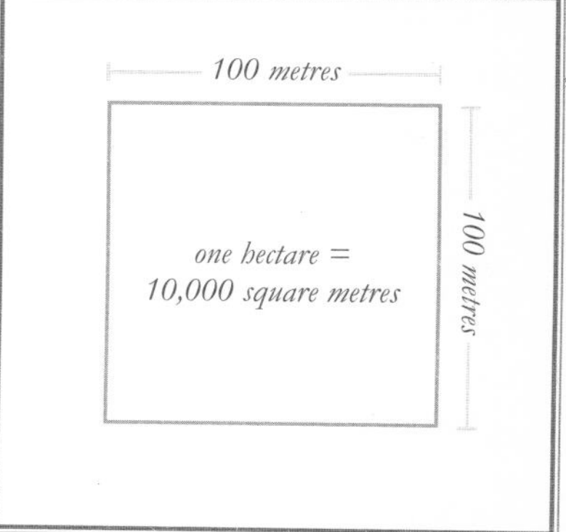

Defining Paper Sizes

In the United States and Canada, there is a complex and historically convoluted system for defining the sizes of standard sheets of paper – a system that includes such sizes as Letter, Legal, and Ledger. Virtually every other country in the world has adopted the ISO paper-size system, and with good reason.

The ISO System

The ratio between the short side and the long side of a rectangle is called its 'aspect ratio', and in the North American system there is no uniformity to the aspect ratio. In the ISO system, however, the aspect ratio is always 1:1.4142, which is the square root of 2, the same relationship that is found between the side of a square and its diagonal.

The big advantage of this system – and one that was first pointed out in eighteenth-century Germany – is that if you cut any such piece of paper in half parallel to its short side, you end up with two pieces of paper that have the same aspect ratio as the original, and this is actually how the different sizes in the series are formed. The width of an A1 sheet is half the height of an A0 sheet. Among other things, this means that when a sheet is enlarged or reduced on a photocopier (for example, from A4 to A3 or vice versa), the copy retains the exact proportions of the original.

If you measure the sides of a sheet of paper in the ISO system, you will find that they do not measure an exact

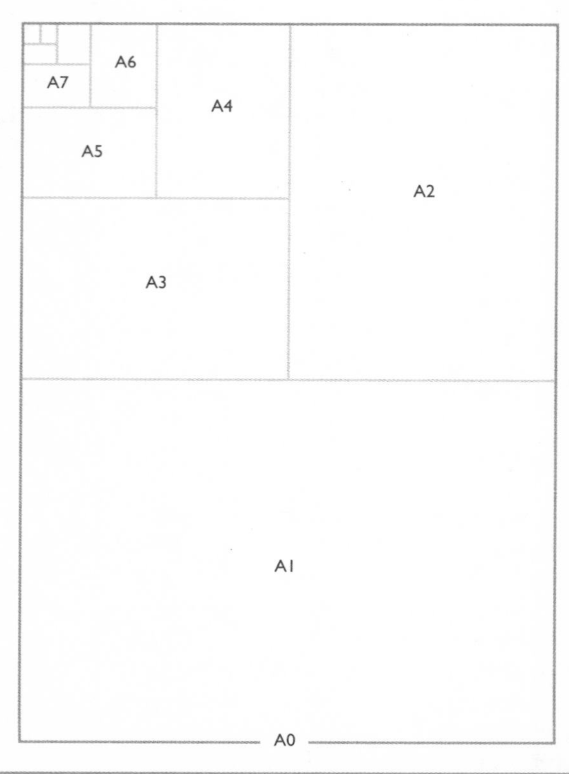

aspect ratio 1: 1.4142

number of millimetres, which might make you wonder whether they are metric at all. However, their sizes are based on a metrically defined area. As well as having an aspect ratio of 1:1.4142, an A0 sheet of paper has an area of exactly one square metre. A1 is half a square metre, A2 a quarter, and so on.

The difference between successive sizes is considerable, but if the ISO A series doesn't meet your need for a specific application, such as a book, the ISO B series provides sizes in between those in the A series. Envelopes that fit folded A and B sizes constitute the C series.

International A Sizes of Paper:

Size	*mm*	*approx. in*
4A0	1682 × 2378	66 1/4 × 93 3/8
2A0	1189 × 1682	46 3/4 × 66 1/4
A0*	841 × 1189	33 1/8 × 46 3/4
A1	594 × 841	23 3/8 × 33 1/8
A2	420 × 594	16 1/2 × 23 3/8
A3	297 × 420	11 11/16 × 16 1/2
A4	210 × 297	8 1/4 × 11 11/16
A5	148 × 210	5 7/8 × 8 1/4
A6	105 × 148	4 1/8 × 5 7/8
A7	74 × 105	2 7/8 × 4 1/8
A8	52 × 74	2 × 2 7/8
A9	37 × 52	1 1/2 × 2
A10	26 × 37	1 × 1 1/2

* *1m²*

U.S. Paper Sizes

Perhaps in recognition of some of the advantages of the ISO paper sizes, the American National Standards Institute (ANSI) defined a new paper standard in 1995. One particular characteristic that this shares with its metric equivalent is that sheets folded along the shorter length will create the next size down. Unlike ISO, however, the aspect ratio does not remain constant.

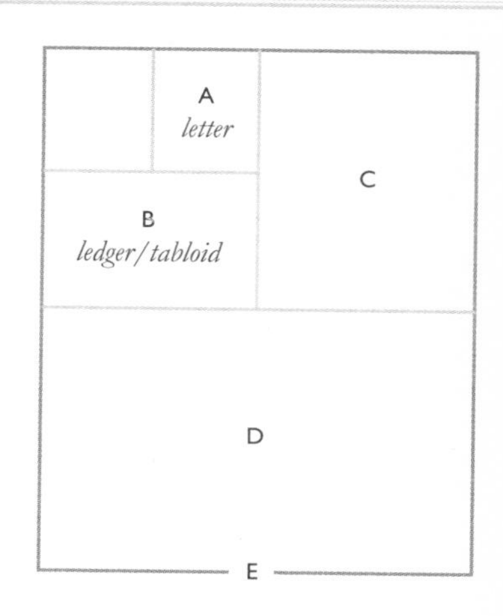

ANSI Sizes of Paper:

Name	*Inches*	*mm*	*ISO**
ANSI A	11 × 8 1/2	279.4 × 215.9	A4
ANSI B	17 × 11	431.8 × 279.4	A3
ANSI C	22 × 17	538.8 × 431.8	A2
ANSI D	34 × 22	863.6 × 538.8	A1
ANSI E	44 × 34	1117.6 × 863.6	A0

* *ISO equivalents are only approximately similar*

CHAPTER FOUR

a measure of volume and capacity

Whether we need to know how much space an object or a quantity of material occupies (its volume) or what quantity of liquid or other material a container will hold (its capacity), measuring in three dimensions is vital to commerce, construction, industry, and science.

'You cannot get a quart into a pint pot.'

Proverb

Metric Units of Volume

As in the case of going from linear to square units, volume can be measured by adding yet another dimension to the square units and, in this case, creating the cubic unit.

The Base Unit

The International System (SI) basic metric unit of area is the linear unit – the metre – squared (see page 48). Likewise, for the unit of volume it is, unsurprisingly, the metre cubed, or cubic metre. The symbol for this is m^3 or, less formally, cu m.

As is the case for metres and square metres, cubic metres can be multiplied and divided by powers of ten to form the derived units such as cubic decimetres (dm^3) and cubic kilometres (km^3).

Area is proportional to the square of the linear dimensions, but volume is proportional to the cube of the linear dimensions. So if all linear dimensions are doubled, the volume becomes eight (2^3) times as great. For this reason, the power of each of the cubic units is three times that of the linear unit with the same prefix. A decametre is 10 metres, but a cubic decametre is 1,000 (i.e. 10 x 10 x 10, or 10^3) cubic metres. A kilometre is 1,000 metres (1 metre x 10^3), but a cubic kilometre is 1,000,000,000 (1,000 x 1,000 x 1,000, or 10^9) cubic metres.

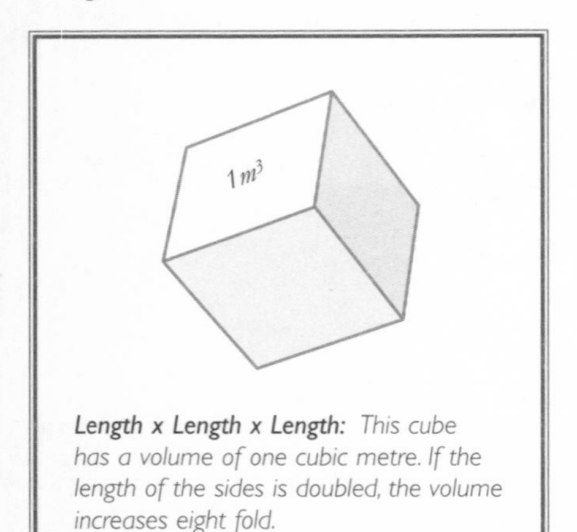

Length x Length x Length: *This cube has a volume of one cubic metre. If the length of the sides is doubled, the volume increases eight fold.*

Litres

A distinction is sometimes made between volume and capacity, the first being used to mean the amount of space that an object occupies and the second referring to the internal volume of a vessel or container. In the SI system, the cubic metre and its derivatives are used in both contexts, but the litre, which was a base unit in an earlier metric system, is still found in common usage as a unit of capacity referring to liquids.

Though not an official SI unit, the litre can be used with SI units and with some SI prefixes (see table). The symbols l and L are both used for the litre, but the capital L is gaining in popularity, as it is less likely to be mistaken for the number 1.

Changing Definitions

As a base unit in the French metric system introduced in 1793, a litre was originally defined as a cubic decimetre – i.e. the volume of a cube with sides one-tenth of a metre (10 cm) long.

SI prefixes for use with litre:

Multiple	*Name*	*Symbols*	*Multiple*	*Name*	*Symbols*
10^0	litre	l L			
10^1	decalitre	dal daL	10^{-1}	decilitre	dl dL
10^2	hectolitre	hl hL	10^{-2}	centilitre	cl cL
10^3	kilolitre	kl kL	10^{-3}	millilitre	ml mL
10^6	megalitre	Ml ML	10^{-6}	microlitre	μl μL
10^9	gigalitre	Gl GL	10^{-9}	nanolitre	nl nL
10^{12}	teralitre	Tl TL	10^{-12}	picolitre	pl pL
10^{15}	petalitre	Pl PL	10^{-15}	femtolitre	fl fL
10^{18}	exalitre	El EL	10^{-18}	attolitre	al aL

At the start of the twentieth century, the definition was changed to be the volume that 1 kg of water would occupy at standard temperature and pressure.

In 1964, the original metre-related definition was resurrected, and the litre has since been defined as a special name for a cubic decimetre ($1\ L = 1\ dm^2$), or the equivalent of one-thousandth of a cubic metre ($1\ L = 0.001\ m^2$).

As a consequence of this redefinition, a litre of water no longer has a mass of exactly 1 kilogram. It is, however, extremely close, and so this remains a useful rule of thumb. By the same token, the mass of 1 millilitre (which is the same as a cubic centimetre) of water is about 1 gram, and 1,000 litres of water have a mass of about 1,000 kg (1 tonne).

Sydharb

Australia has adopted the SI system thoroughly and wholeheartedly, but in a document entitled 'Australian Conventional Units of Measurement in Water', the Australian Water Association includes the 'sydharb' in its list of units. One sydharb is the equivalent of approximately 500 gigalitres (500,000,000 cubic metres), the estimated volume of Sydney Harbour. Australia's total annual water consumption is roughly 48 sydharbs.

Bottle Designations

Given its mystique and the special place that it occupies in the hearts of bon viveurs worldwide, one could almost predict that Champagne would have special names for its bottles. No other form of alcohol is privileged to have quite such a range of size possibilities (although red wine does almost as well).

The glass wine bottle came into use in the late 1600s; in Britain at that time, its capacity was one-fifth of a gallon. The metrically rounded 75 cL is now the accepted international wine-bottle capacity, although the term 'a fifth' is sometimes still applied. Champagne is also available in quarter-bottles (piccolo) and half-bottles (demiboite or fillette). It is as the sizes go up that the naming starts to adopt mythic proportions.

Magnum

Despite its Clint Eastwood connotations, magnum just means 'big', and a magnum of Champagne has a volume of 1.5 L (2 bottles).

Marie-Jeanne

Used only for red wine from Bordeaux in France, a Marie-Jeanne contains 2.25 L (3 bottles).

Jeroboam

The first King of Israel, who led his people to secede from Solomon's kingdom, gives his name to the 3 L bottle, 4 times the volume of the standard bottle. This applies to both Champagne and Burgundy wine. A 3 L bottle of Bordeaux is called a Double Magnum, and a Bordeaux Jeroboam contains 4.5 L.

Rehoboam

It was from the oppression of Rehoboam, son of Solomon and King of Judah, that Jeroboam led the people of Israel. A Rehoboam of Champagne or Burgundy contains 4.5 L (6 bottles). When used to refer to Bordeaux claret, the term Rehoboam means a double Jeroboam, and it contains 9 L.

Methuselah

'And all the days of Methuselah were nine hundred sixty and nine years: and he died.' So the Book of Genesis tells us. The 6 L bottle (8 standard bottles) is named after him, the oldest person on record. The term is used for Champagne and Burgundy, but a 6 L bottle of Bordeaux is called an Impériale.

Bottle of Bubbly: *Champagne bottles range from a glassful to a wedding's worth, and each has its own name.*

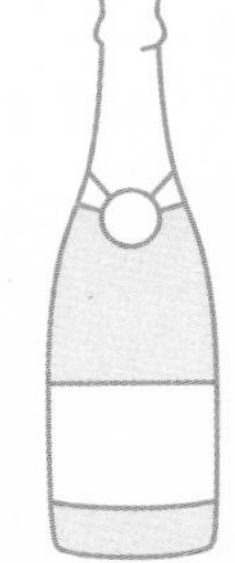

Champagne Bottle Names and Capacities

Type	*Capacity*	*Number of standard bottles*
Piccolo	18.75 cL	1/4
Fillette	37.5 cL	1/2
Bottle	75 cL	1
Magnum	1.5 L	2
Marie-Jeanne	2.25 L	3
Jeroboam	3 L	4
Rehoboam	4.5 L	6
Methuselah	6 L	8
Salmanazar	9 L	12
Balthazar	12 L	16
Nebuchadnezzar	15 L	20
Melchior	18 L	24
Solomon	21 L	28
Sovereign	25 L	33.33
Primat	27 L	36

Salmanazar

The king of Assyria (859–824 B.C.) is remembered in the 9 L (12-bottle) vessel.

Balthazar

The 12 L (16-bottle) Balthazar may recall the Regent of Babylon, son of Nabonidus, or may refer to one of the three wise men who brought gifts to the infant Jesus. They are not named in the Bible, but in the West they have been known as Caspar, Melchior, and Balthazar since the seventh century.

Nebuchadnezzar

Nebuchadnezzar II, King of Babylon in the sixth century B.C., is famous for his feasts, military conquests, for constructing the Hanging Gardens of Babylon, and for the eponymous 15 L bottle.

Melchior

Another member of the three wise men is remembered in the 18 L bottle, but this size and those larger than this are rare – mainly because they are almost impossible to lift.

Solomon

Known as a Salomon in France, the Solomon contains 21 L (28 bottles).

Sovereign and Primat

The 25 L Sovereign (33.33 bottles) and the 27 L (36-bottle) Primat are very rare indeed. Weighing over 140 lb (65 kilos), the Primat is difficult to pour elegantly.

Imperial Units of Volume

The cubic inch, cubic foot, cubic yard, and so on are fairly simple, and these are indeed the units in common use in certain areas of life: the capacity of car engines in the U.S. is generally given in cubic inches, for example.

These units are the same in both the U.S. Customary system and British Imperial system. However, the units used traditionally for measuring out commodities such as liquids and dry goods are far less straightforward.

Gallons, Pecks, and Bushels

The first standardised units of volume in use in England date from the Anglo-Saxon period, when, given the difficulty of measuring the exact capacity of a vessel in cubic units, a unit of capacity was defined by the volume of a specified weight of a known substance. The volume of eight pounds of wheat, for example, was taken as the definition of the gallon, and this formed the basis of the 'Winchester measure' used in medieval England. Winchester was the capital of Wessex, and of England in the tenth and eleventh centuries, and it was here that standards approved by the Crown were kept.

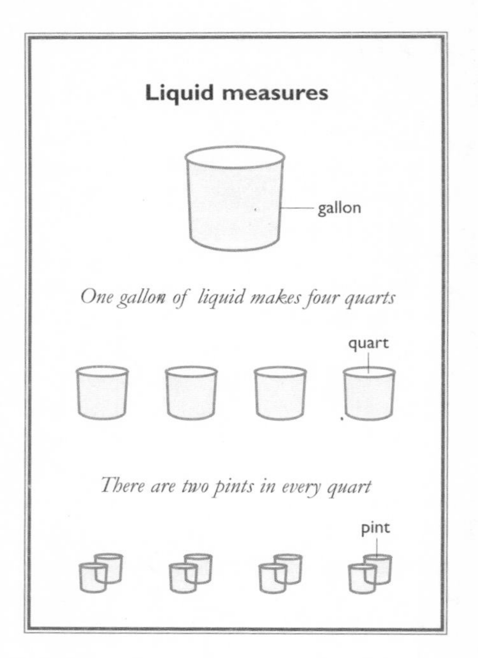

Winchester gallon Winchester pint

The gallon was, and still is, divided into four quarts of two pints each. For dry goods, there were two gallons in a peck, and four pecks in a bushel. Each of these terms was a container of a standard size. For liquid volumes greater than a gallon, a system of barrel sizes that were multiples of the gallon developed, but these were founded on gallons that were defined on the basis of the volume of substances other than wheat (for example, wine or beer), giving rise to alternative definitions of the units of volume.

Two gallons of dry goods make one peck and four pecks make one bushel (not shown)

Narrowing Down the Options

In 1696, the English Parliament defined the Winchester bushel as the volume of a cylinder with specific internal dimensions (that is, 18 1/2 inches in diameter and 8 inches deep). The Winchester gallon, also known as the 'corn gallon', was one-eighth of this bushel and had a volume of 268 4/5 cubic inches. It was used in the measurement of dry commodities.

For liquids, several different gallons remained in use (there were more than a dozen different sizes of dry and liquid gallon in use between 1066 and the end of the seventeenth century), and these included the wine gallon and the ale gallon. In 1706, during the reign of Queen Anne, the wine gallon was defined as having a volume of 231 cubic inches. The ale gallon was considerably larger, at 282 cubic inches.

The Perch

As a linear unit of measurement, the perch is the same as a rod, equalling 16 1/2 feet, but it is also used as a unit of volume. A perch of stonework, for instance, is 16 1/2 feet by one foot by 1 1/2 feet, or 24 3/4 cubic feet (0.70 cubic metre).

Separate Ways

There was a huge range of units of volume in use throughout the British Empire by the eighteenth century. When Britain and the Thirteen Colonies went their separate ways, the United States chose to rationalise the situation by adopting the two most common gallons: the Winchester gallon for dry goods, and the Queen Anne wine gallon for liquids. This meant that a gallon of dry goods occupied about 16 percent more volume than a gallon of liquid.

Unfortunately, Britain made a different decision, opting to use just one gallon for both dry goods and liquids. In 1824, the British Imperial gallon was defined (rather as the litre had been defined by reference to the kilogram) as the volume occupied by 10 pounds of water at specified temperature and pressure. This gave the Imperial gallon a volume of 277.42 cubic inches – much closer to the old ale gallon than the wine gallon. The pint, quart, and bushel were then all redefined on the basis of this gallon.

As a result, the British Imperial units of volume used in the measurement of liquids prior to metrication were about 20 percent larger than those in use in the U.S. (with the exception of the fluid ounce; see page 62).

Liquid Barrel Volumes

Many units of volume are centred around the barrel, primarily because they were developed by the brewing industry as a way of measuring beer and wine.

Historically, the use of 'barrel' as a unit of measurement to define volume comes from actual barrels of certain standard sizes. A traditional barrel is a hollow cylindrical container made of wooden staves and bound with iron hoops. They often have a convex shape, with a slight bulge in the middle that enables the barrel to be rolled on its side with little effort and helps to create a stronger structure.

As with many devices still utilised today, the Ancient Romans were the first people to make commercial use of the barrel – a practice dating from the third century. However, there is evidence to suggest that the Gauls had already been making barrels for several centuries before the Romans.

For nearly 2,000 years, barrels formed the primary method for shipping and storing both liquids and bulk items such as nails and coins. In the late twentieth century, however, barrels slowly lost their importance following the introduction of larger containers, such as the 55-gallon steel drum.

Types of Barrel

Although it is common to refer to barrel-like containers of any size as barrels, this term is technically correct only if the container holds certain precise volumes. In the U.K., a standard beer barrel is 36 U.K. gallons (164 litres), while in the U.S., a standard barrel for liquids is 31 1/2 U.S. gallons (119 litres), or half a 'hogshead'. However, a standard beer barrel is 31 U.S. gallons (117 litres) and a standard dry barrel is equivalent to 105 dry quarts (115 litres).

Volume Equivalents: *Units of volume, using the 119-litre U.S. liquid barrel (E) as the base point. A pint is 1/252 the volume of this barrel, while a hogshead is double its size.*

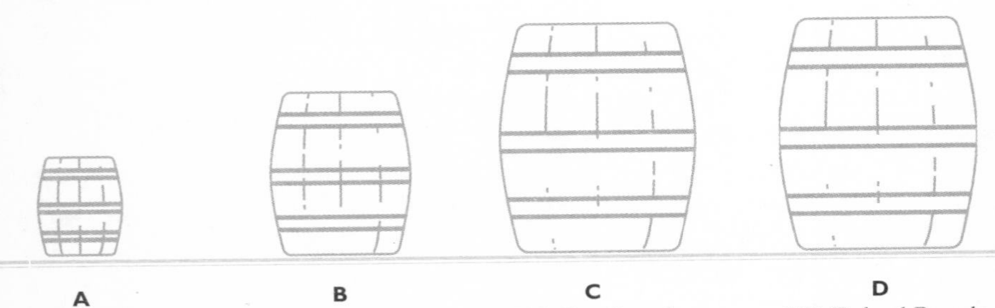

A
Pint (U.S. liquid)
252 per barrel
0.47 L

B
Quart (U.S. liquid)
126 per barrel
0.94 L

C
U.S. Dry Barrel
1.031 per barrel
115.6 L

D
U.S. Federal Barrel
1.016 per barrel
117.3 L

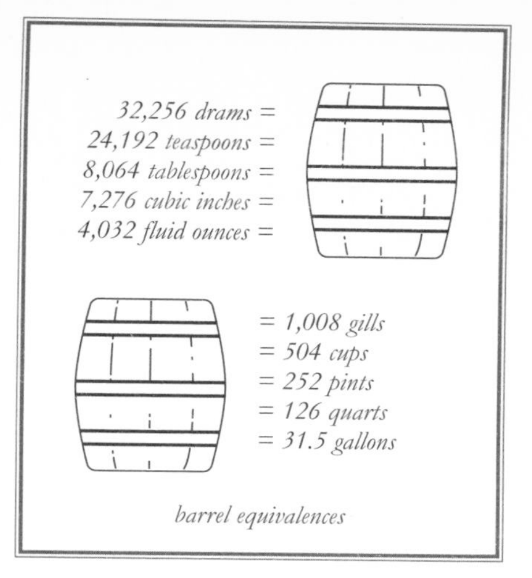

barrel equivalences

The often-misused terms 'keg' and 'cask' refer to containers of any size, the distinction between these being that kegs are used for beers served using external gas cylinders. Ales that undergo part of the fermentation process in their containers are called 'cask ales'.

Oil Barrels

Petroleum products such as crude oil are sold in a standard barrel size (abbreviated bbl) that is equivalent to 42 U.S. gallons, 35 Imperial gallons, or 158.97 litres. Based on an archaic unit of measuring wine – the 'tierce' – this measurement originated in the oil fields of Pennsylvania during the early days of the U.S. oil industry as a way of standardising trade between British and American merchants.

Prior to the introduction of a standard petroleum barrel, oil was commonly sold in 40-gallon (150-litre) whiskey barrels. And even earlier than that, in the late 1850s, Edwin Drake, owner of the first oil well in Titusville, Pa., used washtubs for storage. Only when production started to exceed expectations did he switch to using wooden barrels.

Of course, oil is no longer shipped in barrels at all, but instead is stored directly in the hulls of the tankers (unless it can be transported directly through oil pipelines). Modern oil tankers are classified according to their capacity. A VLCC (Very Large Crude Carrier) typically measures over 200,000 metric tonnes, while a ULCC (Ultra Large Crude Carrier) measures over 300,000 metric tonnes. The largest of all, the *Knock Nevis*, weighs in at 647,955 metric tonnes when fully laden.

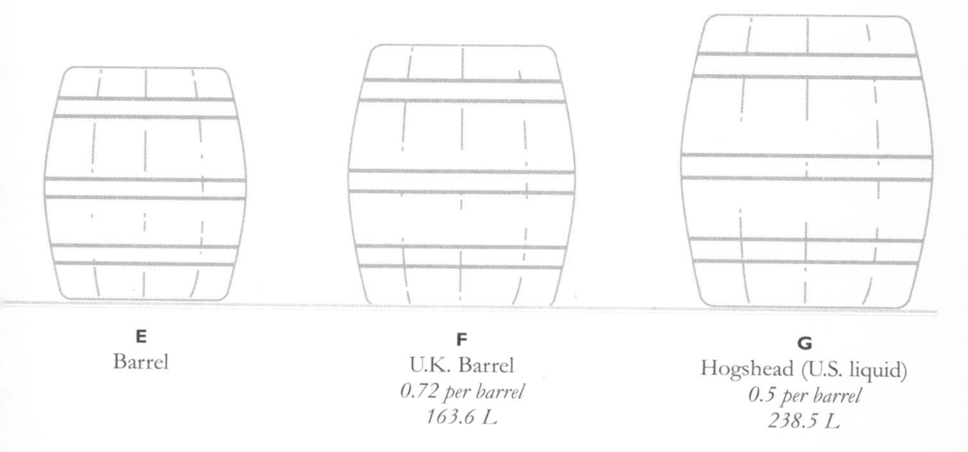

E
Barrel

F
U.K. Barrel
0.72 per barrel
163.6 L

G
Hogshead (U.S. liquid)
0.5 per barrel
238.5 L

Cooking Measurements

When units of mass were used for solid ingredients in recipes, there was little confusion in converting between Imperial, U.S., and metric units. Imperial and U.S. systems both use pounds and ounces, and the metric conversion is straightforward. However, the units of volume used for liquids are now being used increasingly for solids, too, and conversions between these are more complicated.

Pints and Fluid Ounces

We have seen that the U.S. liquid gallon and the U.K. Imperial gallon are different. In fact, the Imperial gallon is more than 20 percent larger. Using their metric equivalents in order to compare them:

1 U.S. liquid gallon = 3.785 L

1 Imperial gallon = 4.546 L

Tablespoons

The tablespoon and its various subdivisions also suffer from a mixed heritage. In the U.S., a tablespoon is defined as half a fluid ounce (about 14.79 mL), whereas countries that use the SI units have a 15 mL tablespoon (except for Australia, where it is 20 mL). The teaspoon is one-third of a tablespoon, making it 4.95 mL in the U.S., and 5 mL in SI countries – including Australia, where there are four teaspoons to the tablespoon, instead of three.

Attempts to define the smallest quantities of liquids used in cooking lead to accuracies of within 1/100 mL, which are often absurd in the context of the kitchen. If this degree of precision is really necessary, it is possible to buy small measuring spoons of the relevant quantities:

Dash: 1/8 teaspoon
Pinch: 1/16 teaspoon
Smidgen: 1/32 teaspoon
Drop: 1/72 teaspoon

Cooking Measures: *For relatively small quantities of ingredients, such as herbs and spices, a set of spoons provides an accurate means of measuring.*

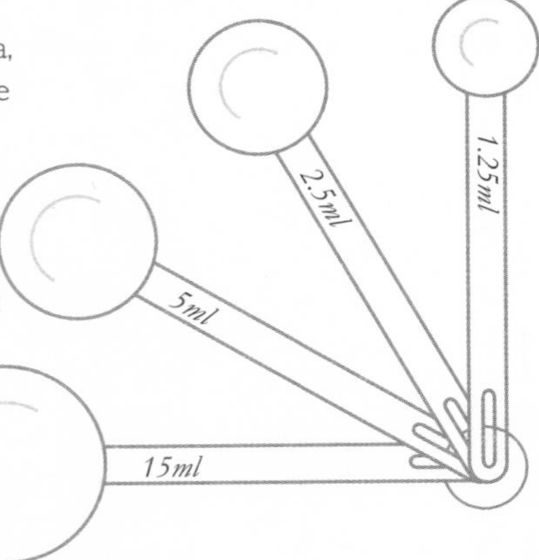

The Weight of Water

Although the Imperial gallon has since been redefined by reference to metric units, the Imperial fluid ounce still weighs almost exactly one ounce avoirdupois. The U.S. fluid ounce is only fractionally heavier, so the old adage holds true:

A pint's a pound the whole world round,
except in Britain, where
a pint of water's a pound and a quarter.

One consequence of this difference between the U.S. and U.K. measurements is that both the pint and the fluid ounce differ in the two systems. The Imperial pint is almost 25 percent larger than the U.S. pint, but in the case of the fluid ounce the difference is not as great as might be expected.

In the U.S., the pint (which is 1/8 of a U.S. gallon) was divided, as the avoirdupois pound is, into 16 ounces – called fluid ounces. The U.S. fl. oz. is therefore 3.785/(8 × 16), which is about 29.6 mL.

The Imperial gallon, on the other hand, was defined as the volume occupied by 10 pounds avoirdupois of water. A pint therefore weighs 10/8 of a pound, or 20 ounces, so the Imperial pint was divided into 20 fluid ounces. An Imperial fluid ounce is therefore 1/(8 × 20) of a gallon. Dividing 4.54609 L by 160 we get about 28.4 mL, which is close to the U.S. fluid ounce. Just over one millilitre per fluid ounce is unlikely to have a major impact on the average recipe, but if the recipe uses pints, you'd better know in which country the recipe book was published.

Cups

In the U.S., where it has been in use for much longer than it has in other countries, the cup is defined as half a U.S. pint, making it about 236 mL. The cup as a unit in cooking has been adopted to some extent in the U.K., also as the equivalent of half a pint – but half a U.K. pint. This makes it about 284 mL. In order to rationalise the situation, Australia has helpfully defined the cup metrically as 250 mL, so there are now three definitions. In Canada, you may also come across the 227 mL cup, which is 8 Imperial fluid ounces, and not a half of any kind of pint.

The problem with using volumes such as cups to define quantities of solid materials can be defined using one word: compressibility. A cupful of shredded lettuce, for example, is an extremely variable quantity. To overcome this, some recipes specify whether the ingredient should be firmly or lightly packed. Some also state whether the cup should be level, rounded, or heaped.

CHAPTER FIVE

a measure of **mass**

The amount of matter in an object has been central to fair trading from the earliest times, and weights for use in a balance have been found in archeological sites 5,000 years old. Today, scientists have the means and the units to determine the mass of objects from subatomic particles to newly discovered stars.

'We are bits of stellar matter that got cold by accident, bits of a star gone wrong.'

Sir Arthur Eddington, Astrophysicist

Mass vs. Weight

There is a common confusion about the terms 'mass' and 'weight' – are they the same, or are they totally different things? The root cause of this is that the term 'weight' has two meanings.

Weight as Mass

The first meaning of weight, and by many centuries the older, is the same as mass: the amount of matter that an object contains. Mass is a measure of the inertia that an object displays, and it is an unchanging quality. Inertia is resistance to acceleration – the greater the mass of an object, the more force is required to speed it up, slow it down, or change the direction in which it is moving. The mass of an object is the same in all locations and under all conditions. If you are travelling in a rocket through space under conditions of weightlessness, or zero gravity, a cannonball will have no weight (see below), but if you try to hurl it across the capsule it will take exactly the same effort to do so as it would on earth. It has the same mass – the same resistance to being accelerated – regardless of where it is. When we talk about the weight of an object or a quantity of matter, we almost always mean its mass.

Weight as Force

However, the term weight does have a very specific meaning in some scientific contexts, when it means the force resulting from the action of gravity on a body. The weight of an object in this sense is the product of the object's mass and its acceleration due to gravity.

F (force or weight) = m (mass) x a (acceleration due to gravity).

The weight of an object therefore changes as gravity changes. Thus, the weight of an object on the moon is less than it is on earth, even though its mass is the same, because the moon has less mass than the earth and the gravitational pull that the moon exercises on the object is therefore less (see page 112).

Gravity and Weight

'From the moment they had left the Earth, their own weight, and that of the Projectile and the objects therein contained, had been undergoing a progressive diminution [. . .] Of course, it is quite clear, that this decrease could not be indicated by an ordinary scales, as the weight to balance the object would have lost precisely as much as the object itself. But a spring balance, for instance, in which the tension of the coil is independent of attraction, would have readily given the exact equivalent of the loss.' (From *Round the Moon* by Jules Verne)

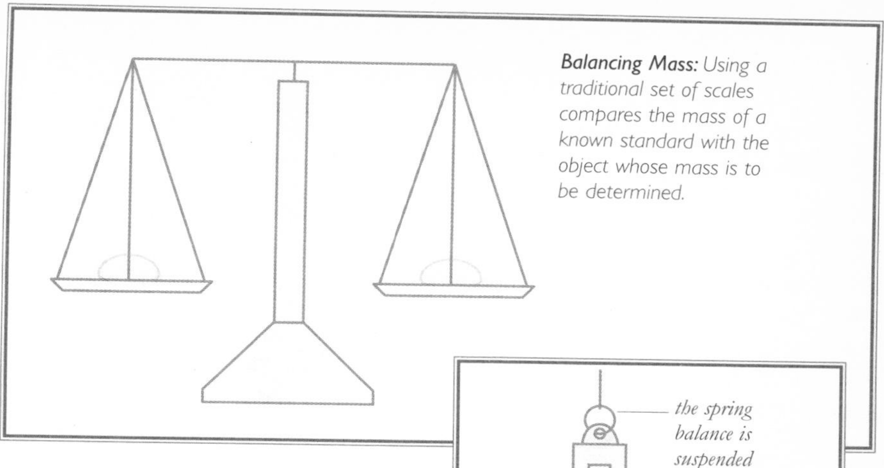

Balancing Mass: *Using a traditional set of scales compares the mass of a known standard with the object whose mass is to be determined.*

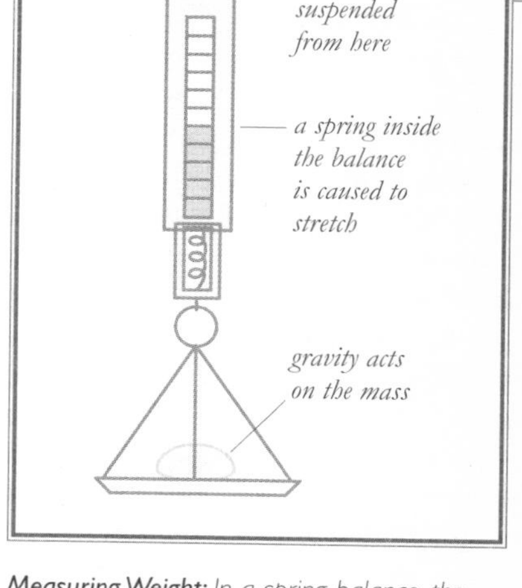

Measuring Weight: *In a spring balance, the extension of a spring indicates the effect of gravity on the unknown mass, in other words, its weight.*

Measuring Weight

Since ancient times, weighing has involved using scales to compare an unknown object or quantity with a known standard. This process compares two masses – when the scales are balanced, the forces acting down on the two ends must be the same. We know that the acceleration due to gravity is the same at each end, so the masses of the two objects must be the same. By establishing the mass of the object in reference to a standard mass, this method of weighing is independent of gravity and is therefore equally effective on earth or on the moon.

Not until the eighteenth century, once it was established that the extension of a spring was in direct proportion to the force exerted upon it, was it possible to measure the weight of an object as the force that it exerts due to gravity. A spring balance would indeed give a different reading if the same object were weighed on the moon. However, this doesn't mean that we can't use it to measure mass – we just need to make sure that it is correctly calibrated in the gravitational conditions in which it is being used. This is why the 'weights and measures' inspector examines it periodically by checking the weight reading that it gives when a standard mass is placed upon it.

In this chapter, we will be looking at units that are used to measure mass. Weight, in the sense of a force, will be covered in Chapter 9.

Imperial and U.S. Customary Units of Mass

Both the Ancient Greeks and the Romans had units of mass that are close to the modern pound. That of the Greeks weighed about 0.94 pounds, while that of the Romans was somewhat lighter, at about 0.72 pounds.

The Pound

Over the centuries there have been several versions of the pound in Britain, including the troy pound, the apothecaries' pound, the wool pound, and the Tower or merchants' pound. All of these went out of use before the implementation of the Imperial system, in which the units of mass are based on the avoirdupois pound.

Units of weight in the U.S. Customary system are also based on this pound. The basic unit of the pound is the grain, originally based on the weight of a grain of wheat. (In SI units, a grain weighs 0.064798 grams.) The avoirdupois pound comprises 7,000 grains, and is divided into 16 ounces, each weighing 437.5 grains.

In the pre-SI British system, the units of mass were:

16 drams	=	1 ounce
16 ounces	=	1 pound
7 pounds	=	1 clove
14 pounds	=	1 stone
28 pounds	=	1 tod/quarter
112 pounds	=	1 hundredweight
364 pounds	=	1 sack
2240 pounds	=	1 ton
2 stones	=	1 quarter
4 quarters	=	1 hundredweight
20 hundredweight	=	1 ton

Cloves, quarters, tods, and sacks went out of use in Britain long before the introduction of decimal units, but pounds and stones are still commonly used, especially when referring to a person's body weight. The sack is a peculiar weight, not being divisible by the hundredweight, nor dividing into the ton.

Hundredweights and Tons

From the list above it can be seen that the stone is a key unit: 2 stones make a quarter or tod; 8 stones make a hundredweight; and 26 stones make a sack. There is good reason to suppose that this is why the hundredweight and the ton consist of such curious numbers of pounds, since one would assume that a hundredweight should be 100 pounds and a ton would then be 2,000 pounds. It has been suggested that the hundredweight (abbreviated to cwt) came about in the coal-mining industry, and that the extra 12 pounds accounted for unburnable rock and slag in the coal. This gave rise to the gross, or 'long', cwt of 112 lb and the net, or 'short', cwt of 100 lb.

The Slug

As we saw on page 66, one definition of mass is as a measure of inertia. Whereas most units of mass are also used as units of weight, which is a force, the 'slug' is specifically a unit of mass in the English foot–pound–second system. Now rarely used, the unit was devised by the English physicist A. M. Worthington in the early 1900s, who defined it as: 'the mass in which an acceleration of one foot per second per second is produced by a force of one pound.' Since acceleration due to gravity is 32.17404 feet per second per second in English units, the slug is equal to 32.17404 pounds.

U.S. Customary Units

The ounces and pounds used in the U.S. Customary system are the same avoirdupois units that are found in the Imperial system, but stones didn't make their way across the Atlantic. There was therefore no impediment to making a hundredweight 100 lb and a ton 2,000 lb, and we find both the long ton (2,240 lb) and the short ton (2,000 lb) in use in the U.S. The term 'ton' on its own is usually taken to mean a short ton. The SI tonne (1,000 kg) is very close to the long ton, and is called a 'metric ton' in the U.S. and U.K.

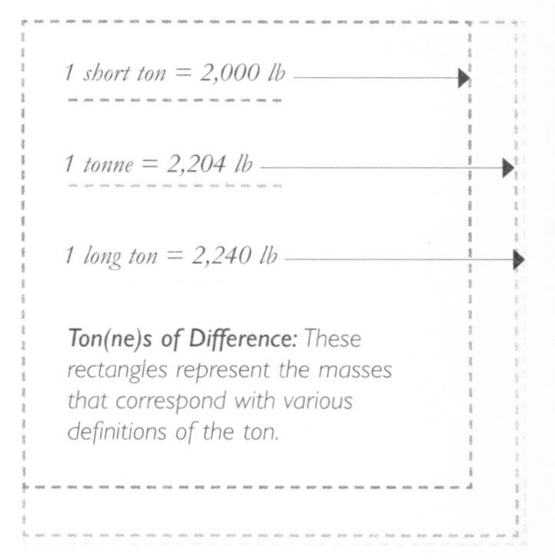

Ton(ne)s of Difference: *These rectangles represent the masses that correspond with various definitions of the ton.*

U.S. Customary Units

Unit	*Divisions*	*SI Equivalent*
1 grain (gr)	1/7000 lb	64.79891 mg
1 dram (dr)	27 11/32 gr	1.7718451953125 g
1 ounce (oz)	16 dr	28.349523125 g
1 pound (lb)	16 oz	453.59237 g
1 (short) hundredweight	100 lb	45.359237 kg
1 (long) hundredweight	112 lb	50.802 kg
1 (short) ton	2,000 lb	907.18474 kg
1 metric ton	2,204.6 lb	1,000 kg (1 tonne)
1 (long) ton	2,240 lb	1,016 kg

Small Amounts and Precious Goods

Historically, separate systems grew up in Europe to define the masses of various kinds of substance. The pounds and ounces that we encounter in the Imperial and U.S. Customary systems are the avoirdupois units used to measure quantities of commercial goods that have a significant mass, but in medieval Europe two other systems evolved that used the same units, only with different values.

The Troy Ounce

The troy system of mass probably originated in Troyes, an important trading city in the nort-east of France not far from Paris. Used specifically to weigh gold and other precious metals, the troy system probably came to Britain with the Norman French at the time of William the Conqueror (1027–87), if not before. Some elements of this system are still in use today.

The units of the troy system are the troy pound, the troy ounce, the pennyweight, and the grain. The grain is the same unit that we find in the Imperial and U.S. systems (the equivalent of 64.8 mg). Given that troy units are for measuring substances with a high value, you might expect that the troy ounce would be a smaller unit than the Imperial ounce, but in fact it is larger. While an Imperial ounce weighs 437.5 grains, the troy ounce weighs 480 – about 10 percent more. The logical extension of this would be for a troy pound to be heavier than an Imperial pound, but it isn't – a troy pound is 12 troy ounces and not 16, making a troy pound 5,760 grains, while an avoirdupois pound is 7,000 grains.

Gemstones and precious metals are still measured in troy ounces and grains, but the pennyweight and troy pound are no longer used.

Old Money

The troy system was at the heart of the pre-decimal monetary system in Britain, introduced by King Henry II (1133–1189). At the time, a penny coin was one 'pennyweight' of silver. A pennyweight was 1/20 of a troy ounce, or 1/240 of a troy pound of sterling silver – or a 'pound sterling', as the unit of currency became known. It seems a little contrary that the pound sterling was composed of 20 units (shillings) of 12 pennyweights (pennies), rather than 12 units of 20 like the troy pound.

Worth its Weight: *The British units of currency introduced by Henry II lasted for 800 years.*

Measuring Medicines

Another system, closely related to the troy system, was devised by the apothecaries. This used both the troy pound and the troy ounce, but the apothecaries' ounce was divided into eight 'drachms' each composed of three 'scruples' – a scruple weighing 20 grains.

Apothecaries had their own symbols to denote the drachm and the scruple. The origin of the word scruple is the Latin *scrupulus*, meaning 'small stone'. Just as now, the Romans also used the word figuratively to mean a source of discomfort, anxiety, or moral unease. If someone doesn't have an ounce of scruples, we can assume that he isn't an apothecary.

In the early nineteenth century, after the introduction of the standard Imperial gallon that related liquid volume to mass, the apothecaries' units of mass were extended to include the 20-ounce pint and the 8-pint gallon. In this version of the system, the 'minim' replaced the grain as 1/20 of a scruple. The apothecaries' system was still being used to measure drugs in the early part of the twentieth century, but pharmacists now use the metric system almost exclusively.

Worth a Bean

Seeds and cereal grains have been used as small units of measurement throughout the ages, and the Ancient Greeks used carob beans, or 'keration'. This gave rise to the carat as a unit for weighing diamonds and other precious stones. In the Imperial and U.S. Customary systems, the carat was defined as 3.2 troy grains (about 207 mg), but all jewellers now use the metric carat, established in 1907 as 200 mg.

Fly-Line Weights

The weight of the line used for fly-fishing needs to be matched to the strength of the rod for effective casting. In 1961, the American Fishing Tackle Manufacturers Association introduced a code that assigned an AFTMA weight rating to lines so that they can be matched to the correct type of rod. The system uses the weight of the first 30 feet of fly line, measured in grains, as the basis of the weight rating. The table below shows the fly-line weight designations and their grain weight.

AFTMA Rating	*Weight (grains)*
1	60
2	80
3	100
4	120
5	140
6	160
7	185
8	210
9	240
10	280
11	330
12	380

In Britain, jewellers also use the carat to describe the purity of gold alloy. This unit is called a karat in the U.S. It represents 1/24 part, so the composition of 18-karat gold is 18/24 (or 75 percent) pure gold.

Metric Mass

Of the seven base units on which the SI system is founded, six are defined in terms of natural phenomena. Only the kilogram is now defined by a physical prototype.

The definition of the kilogram was originally (i.e. when the first French metric system was introduced at the end of the eighteenth century) defined as the mass of a cubic decimetre of water. It was therefore linked to the metre, which was in turn defined by reference to the circumference of the earth. In 1889, however, the first Conférence Générale des Poids et Mesures (CGPM) agreed to the creation of an international prototype of the kilogram, and made this the unit of mass. Made of an alloy of platinum and 10 percent iridium, the prototype still resides at the International Bureau of Weights and Measures under the conditions specified by the first CGPM.

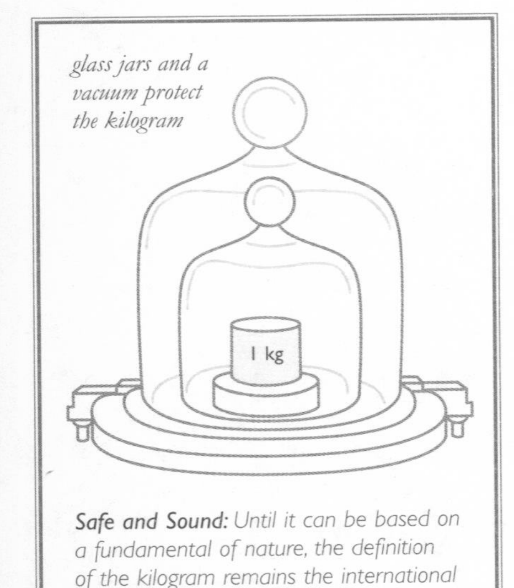

Safe and Sound: *Until it can be based on a fundamental of nature, the definition of the kilogram remains the international prototype, which is stored in Paris. Copies around the world are periodically compared with this.*

Concerning the distinction between mass and weight, the third Conférence Générale stated in 1901, 'The kilogram is the unit of mass [and not of weight or of force]; it is equal to the mass of the international prototype of the kilogram.' The unit of force in the SI system is the newton, and its relationship to mass can be seen in its definition: one newton is the force required to give a mass of one kilogram an acceleration of one metre per second per second (see page 112).

A Better Definition?

Each country that subscribes to the SI system has a copy of the prototype kilogram to use as its standard, but the situation is not perfect. Over time, each copy accumulates dirt or scratches that affect its weight, so each one must be returned periodically to Paris to be cleaned and verified.

An ideal base unit is one that, like the other six SI base units, can be replicated anywhere in the world simply on the basis of information and relatively simple equipment, and there is a worldwide effort in progress to achieve such a definition for the kilogram by relating it to other physical constants. Suggestions so far include defining the kilogram as:

SI Multiples

As with all other SI units, the unit of mass can be scaled up and down by powers of 10 indicated by the prefixes that we have already met, but the unit to which the prefixes are added is not, in this case, the base unit but the gram. A kilogram is, of course, 1,000 grams. The gram was a base unit in the centimetre–gram–second (CGS) system, which was replaced by the metre–kilogram–second system, of which the SI system is an extended version.

Multiple	*Name*	*Symbol*	*Multiple*	*Name*	*Symbol*
10^0	gram	g			
10^1	decagram	dag	10^{-1}	decigram	dg
10^2	hectogram	hg	10^{-2}	centigram	cg
10^3	kilogram	kg	10^{-3}	milligram	mg
10^6	megagram	Mg	10^{-6}	microgram	µg
10^9	gigagram	Gg	10^{-9}	nanogram	ng
10^{12}	teragram	Tg	10^{-12}	picogram	pg
10^{15}	petagram	Pg	10^{-15}	femtogram	fg
10^{18}	exagram	Eg	10^{-18}	attogram	ag
10^{21}	zettagram	Zg	10^{-21}	zeptogram	zg
10^{24}	yottagram	Yg	10^{-24}	yoctogram	yg

- the mass of a fixed number of atoms of silicon or carbon
- a mass that is given a specific acceleration by a certain force
- the mass of a specified number of electron mass units
- the mass of a body at rest whose equivalent energy corresponds to a specific frequency (defining the kilogram in relation to Planck's constant)
- the mass of a superconducting body levitated in a magnetic field generated by a superconducting coil carrying a specific current.

The Metric Ton

A 'megagram' is 1,000,000 grams, or 1,000 kg. In countries that use the SI system, this mass is sometimes referred to as a tonne, or a metric ton, the spelling 'tonne' being derived from the French. The tonne is not an SI unit, but is accepted for use with the SI. The official symbol for tonne is t, and SI prefixes can be added to this to create units such as the megatonne (Mt). The Imperial long ton of 2,240 lb (approx. 1,016 kg) is very close to the weight of a tonne, and many people in Britain make little distinction between the two. In the U.S., where the ton (or short ton) weighs 2,000 lb, the term 'metric ton' is generally used to refer to the tonne.

Mass on the Atomic Scale

The mass of a proton or a neutron is about is 1.67 x 10^{-24} grams, or 1.67 yoctograms, the smallest unit in the SI system of prefixes. But since no device can measure the mass of such tiny particles directly, how is it possible to know?

The starting point is the unified atomic mass unit (u), also known as the dalton (Da). This was previously known as the atomic mass unit (amu), when it was the equivalent of the weight of a hydrogen atom. The hydrogen atom consists of a proton and an electron, but since the weight of the electron is 1/1,836 that of a proton, its contribution is negligible and the atomic mass unit was more or less equal to the mass of a proton or neutron.

This has since been revised to take into account the extra mass that exists in the form of the energy that binds protons together in the nucleus, and the unified atomic mass unit is now defined as 1/12 of the mass of a carbon-12 atom (the nucleus of which consists of six protons and six neutrons).

Weighing an Atom

If the mass of a carbon atom is used as the basis for a unit of measurement, surely this means that its exact weight has to be established first? In fact, it does not, because the unified atomic mass unit is effectively unitless – it is a measure of the relative mass of an atom. The reason this works is that – thanks to the Italian scientist Amedeo Avogadro (1776–1856) – it is possible compare the masses of different elements and compounds.

Amedeo Avogadro
1776–1856

Avogadro put forward the – since proven – hypothesis that equal volumes of gases, at the same temperature and pressure, contain equal numbers of molecules. It is therefore possible to weigh a given volume of hydrogen (which has an atomic mass of approximately 1 u), compare it with the weight of the same volume of oxygen under the same conditions, and work out the atomic mass of oxygen (which proves to be almost exactly 16 u).

Avogadro and the Mole

So how does this lead to an actual value for the mass of a proton? The key here is the 'mole', which is one of the seven SI base units – in this case, the unit for the amount of substance. A mole is the atomic mass of a substance in grams. For example, the atomic mass of carbon is 12, so a mole of carbon is 12 g. A mole of hydrogen is 1 g, and a mole of oxygen is 16 g. The important aspect of this is that a mole always contains the same number of atoms or molecules of that particular substance, and that number is known

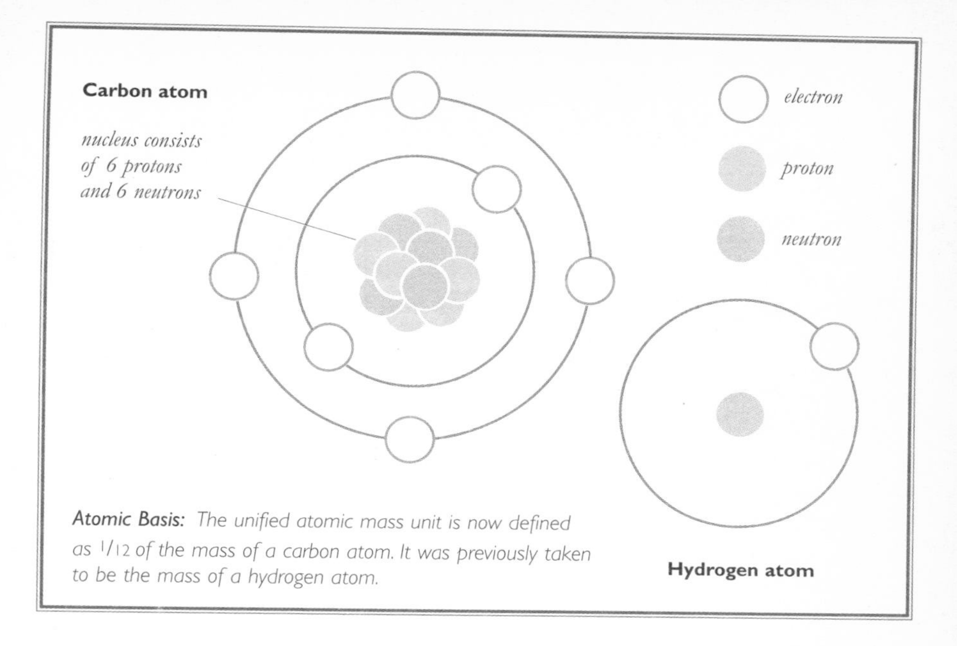

Atomic Basis: The unified atomic mass unit is now defined as 1/12 of the mass of a carbon atom. It was previously taken to be the mass of a hydrogen atom.

as Avogadro's number (named in honour of the scientist, not because he discovered it).

Avogadro's number continues to be refined, but currently has a value of $6.02214199 \times 10^{23}$. In other words, a mole of any element contains more than 60 billion trillion atoms. Given that one gram of hydrogen contains Avogadro's number of hydrogen atoms, and that each hydrogen atom contains one proton, we have the means to work out the exact weight of a proton, which proves to be 1.6726 yg.

Sensitive Scales

Although there is, as yet, no instrument capable of measuring a mass as small as a yoctogram, in 2005 a team of scientists at the California Institute of Technology (Caltech) succeeded in building 'scales' that can detect a cluster of xenon atoms with a mass of a few zeptograms (10^{-21} g). This is approximately the mass of a single protein molecule.

At the heart of the scales is a microscopic blade of silicon carbide, which vibrates in a magnetic field and generates a voltage in a wire. A molecule placed on its surface reduces the frequency of the vibration and alters the voltage, giving a measure of the mass of the molecule. The blade is about 1,000 nanometres long, so if 10,000 of these were placed end to end their total length would be 0.4 inches (1 cm).

The Density of Matter

Everyone is familiar with the riddle 'Which weighs more, a ton of coal or a ton of feathers?' (although, to be precise, the question should be 'Which has the greater mass?'). The answer, of course, is that they both weigh the same. The difference between them is that they occupy different volumes.

Two objects that have the same mass but occupy different volumes are said to have different densities. The density of matter is its mass per unit volume, and this is a product of how tightly the atoms or molecules in the substance are packed.

Units of Density

The Imperial and U.S. Customary units for mass are primarily the ounce and the pound, and those for volume are the cubic inch, cubic foot, and cubic yard. So, the units in which density is expressed are commonly ounces per cubic inch (oz/in^3), pounds per cubic inch (lb/in^3), pounds per cubic foot (lb/ft^3), pounds per cubic yard (lb/yd^3), and (rarely) slugs per cubic foot. The density of a liquid can be expressed in pounds per gallon (lb/gal).

In the SI system, the derived unit of density is the kilogram per cubic metre (kg/m^3). For convenience, other acceptable combinations of SI and SI-compatible units, such as grams per cubic centimetre (g/cm^3) or per millilitre (g/mL) and kilograms per litre (kg/L), are also used.

Archimedes
c.287–212 B.C.

Measuring Density

To measure the density of a solid or liquid, we need to know the mass of a sample and its volume. In the case of a liquid, this is relatively straightforward. Simply transfer a known volume of the liquid into a flask of known mass, weigh the flask with the liquid in it, and subtract the weight of the flask. The mass of the liquid can then be divided by its volume to determine the density.

To find the density of a solid object or material, it is necessary to know its volume. As stated by Archimedes' principle of buoyancy, this can be discovered by submerging the object (or substance) in a flask containing water and seeing how much the level of the water rises – the change in volume of the liquid is exactly the volume of the submerged object. The object can then be weighed and the mass can be divided by the volume to determine the density.

Specific Gravity

In order to make quick comparisons between the densities of different substances, a measurement called 'specific gravity', or 'relative density', is often used. This is an expression of the density of a material relative to a reference material (usually water). A material with a specific gravity of 3.5 is three-and-a-half times as dense as water. Specific gravity has no units.

Hydrostatic Weighing

According to Archimedes' principle, when a body is immersed in a liquid, the upthrust on the object is equal to the weight of the liquid displaced by the object. Using water as the liquid, we can use this fact to determine specific gravity. A sample of the material is weighed in air and is then weighed underwater. The difference between the two weights is the result of the upthrust due to the water that has been displaced (which has precisely the same volume as the sample). The specific gravity of the sample is therefore its weight in air divided by the upthrust of the displaced water. If, for example, the sample had the same density as water, the upthrust would be equal to the weight of the sample in air, giving a specific gravity of one. This method works well for materials with a specific gravity greater than one, but less well for materials that float, as the force required to submerge the sample has to be taken into account.

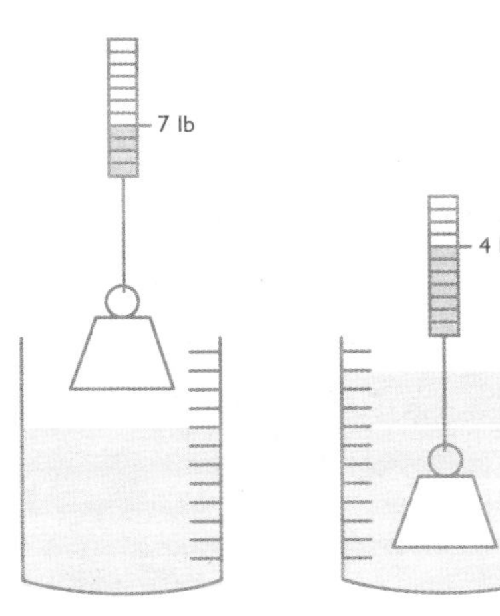

Weight and Density: *Because hydrostatic weighing gives an accurate measure of the density of an object, it can be used to determine the relative fat content of a person's body, and this method is used in health assessment and fitness training.*

In the SI system, the density of water is 1,000 kg/m^3. This is the equivalent of 1 g/cm^3, so a material with a specific gravity of 3.5 has a density of 3.5 g/cm^3. In other words, the specific gravity of a substance is $^1/_{1,000}$ of its density expressed in kg/m^3.

The density of the elements ranges from 0.09 kg/m^3 in the case of liquid hydrogen, to 22,650 kg/m^3 for iridium, meaning that a cubic metre of iridium has a mass of 22.65 tonnes. Lead is only half this dense, and a cubic metre of water has a mass of one tonne. The density of air – 1.29 kg/m^3 – is 14 times greater than that of hydrogen, which is why a hydrogen-filled balloon rises.

CHAPTER SIX

a measure of temperature

Concepts such as length, area, and volume are relatively straightforward and comprehensible, but temperature – and the measurement of it – is rather more complex than it might at first appear.

'It doesn't make a difference what temperature a room is, it's always room temperature.'
Steven Wright, Radio Presenter

Hotness and Coldness

While humans have an intuitive – even instinctive – understanding of heat and cold, and we know that a thermometer measures this quality, we rarely stop to consider what temperature actually is, or how we are able to measure it.

When two objects at different temperatures are placed together, the hotter one will become cooler and the colder one will become warmer until they are at the same temperature. At one time, this was thought to be the result of a flow of 'caloric' – an invisible, intangible liquid – from one to the other, like water flowing from a higher container to a lower container until the levels are in equilibrium.

At the beginning of the nineteenth century, Benjamin Thomson cast doubt on this theory. Observing the heat generated when a cannon barrel was drilled, he suggested that it might be the motion itself that resulted in the change of temperature. Fifty years later, J. P. Joule showed that heat is a form of energy, and that in fact it is energy that 'flows' from a hotter to a cooler object.

Temperature is a measure of the average kinetic energy of the particles in a sample of matter – a measure of how fast the molecules in the material are moving.

Methods of Measurement

In order to measure temperature, it is necessary to use a material whose properties change steadily and in a measurable way as the kinetic energy increases. If it is certain that the material will end up at the same temperature as the object with which it is in contact, it is therefore possible to measure the temperature of that object.

In the early seventeenth century, the expansion of a gas as its temperature rises was found to behave in this linear fashion, and the first thermometers used this principle – the expanding gas pushing a liquid along a tube. The expansion of various liquids was then used, and the first sealed thermometer, which contained alcohol, was made in 1641. Robert Hook later increased the clarity of the reading by adding red dye to the alcohol.

It was Daniel Gabriel Fahrenheit, an instrument maker of Danzig and Amsterdam, who first used mercury as the 'thermometric' liquid. This bright, silvery liquid had several advantages: it is easy to see, expands uniformly, doesn't stick to the glass tube of the thermometer, and remains liquid through a wide range of temperatures.

Expanding the Range

Although alcohol and mercury are the most common materials used today in climatic and medical thermometers, they have their limitations when it comes to more precise industrial and scientific research purposes, not least because they freeze and vaporize at low and high temperatures. Fortunately, there

The Bimetallic Strip *Two metal strips, each with a different coefficient of linear expansion, are joined together so that the strip is straight at a particular reference temperature. Typical combinations are brass and steel, or copper and invar. When the temperature is raised or lowered, and the two expand or contract by different amounts, the bimetallic strip bends.*

brass

steel

at the reference temperature, the strip is straight

as the temperature increases, the brass expands more and bends the strip down

when the strip is cooled, the brass contracts more and the strip bends upward

are several other materials with equally suitable properties that can be used to gauge temperature.

Gases at low pressure all display the same very simple relationship between their pressure, volume, and temperature, and this has made it possible to set up an absolute temperature scale that is independent of the thermometric medium. Temperature defined in this way is called the 'thermodynamic' temperature, and this scale has a naturally defined zero – the point at which the pressure of the gas is zero (i.e. there is no kinetic energy in the gas), making the temperature also zero (see page 84).

Like gases and liquids, metals also expand as the temperature increases, and this property is exploited in the bimetallic strip. If two strips of different metals that expand at different rates are bonded together, a rise in temperature will cause the strip to bend as one side expands more than the other. A bimetallic coil can be used to make a very sensitive thermometer. Since a bimetallic strip will conduct electricity, it can also be used as a temperature-sensitive switch, bending to make or break an electrical contact in a kettle, stove, or central-heating system.

Temperature also affects the electrical resistance of a metallic conductor, and a platinum resistance thermometer can be used to measure temperatures over an enormous range, from about -260 °C to 1,235 °C.

When wires of different metals are fused together, an electrical current flows between them – a current that is proportional to the temperature. 'Thermocouples' can therefore be used as thermometers, and they are commonly found in industrial applications.

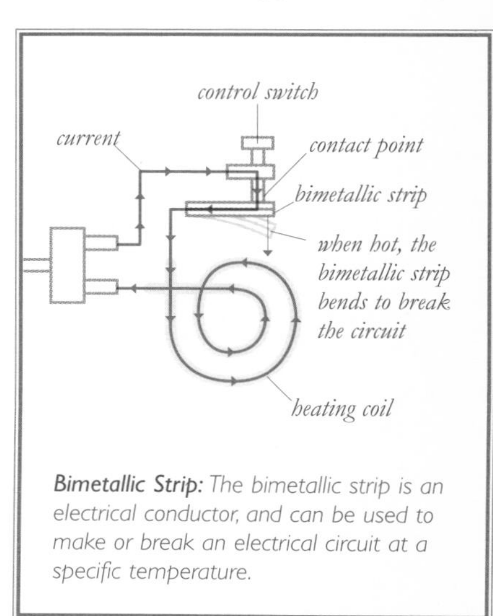

Bimetallic Strip: *The bimetallic strip is an electrical conductor, and can be used to make or break an electrical circuit at a specific temperature.*

A Matter of Degree

Temperature is always measured in degrees, but historically there have been many different values and scales for the degree, and several different choices of reference points for the fixed values on the scale.

The first sealed thermometer had a scale divided into 50 'degrees', but it had no reference point or zero, so the scale was entirely arbitrary. In 1664, Robert Hook used the freezing point of water of as a 'fixed point', and he created a scale in which divisions corresponded to about 1/500 of the volume of the liquid in the thermometer. In 1702, the Danish astronomer Ole Roemer, fixing the temperature of melting ice as 7.5 degrees, added the boiling point of water (which he called 60 degrees) as a second fixed point.

Several years later, the German physicist Daniel Gabriel Fahrenheit visited Roemer and adapted his system. Fahrenheit used three fixed points: the temperature of a mixture of ice, salt, and water; the freezing point of water; and the body temperature of a healthy man. These he fixed as 0, 32, and 96 degrees. The boiling point of water was later fixed at 212 degrees, effectively dividing each of Roemer's degrees into four, and after a few readjustments the Fahrenheit system was in place. It was to remain the temperature standard in most English-speaking countries until

Ole the Polymath

Although he was not the only one to display such qualities in the academic world of his time, Ole Roemer showed remarkably wide-ranging interests and achieved fame in many scientific fields. As well as contributing to the history of the thermometer, he was instrumental in introducing a comprehensive and well-defined system of weights and measures in his native Denmark – including the Danish mile, which, like the British Admiralty nautical mile, was based on minutes of arc of latitude. He was also a noted social reformer, mathematician and astronomer, and was the first to demonstrate that light travelled at a finite speed (which was probably the first universal constant to be demonstrated).

Ole Roemer *1644–1710*

Converting Temperatures

Since the range from freezing to boiling is 180 degrees on the Fahrenheit scale and 100 degrees on the Celsius scale, the Fahrenheit degree equals 100/180 (or 5/9) of a Celsius degree. To convert a temperature from °F to °C, subtract 32 from the Fahrenheit temperature, divide the result by 9, and multiply by 5. To convert from °C to °F, divide by 5, multiply by 9, and add 32.

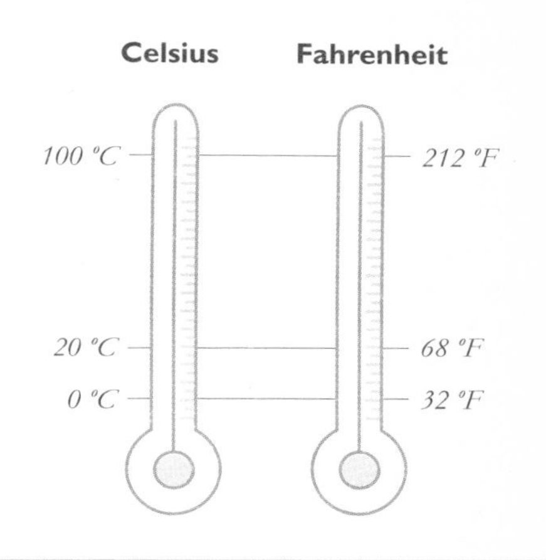

the 1960s, when the Celsius scale was phased in as part of the process of metrication. The Fahrenheit system is still in use in the U.S. and Jamaica.

The Centigrade Scale

A 100-degree scale was proposed by Anders Celsius, a Swedish astronomer, in 1742. He chose the freezing and boiling points of water as his fixed points, but somewhat counter-intuitively made the freezing point 100 and the boiling point zero. In 1745, shortly after the death of Celsius, the botanist Carl Linnaeus (who, incidentally, had been strongly influenced by Anders' botanist uncle Olaf Celsius) took the step of reversing the scale, as this suited his purposes better. He was anxious to observe and control the environment in which his plants were growing, and zero as the freezing point indicated the temperature at which many plants started to suffer.

This 'Centigrade' scale was widely adopted, but because grades and centigrades are also used as measures of angle, it was officially renamed the Celsius scale, using degrees Celsius (°C), in 1947. The Centigrade and Celsius scales differ by a fraction of a degree because the Celsius scale is defined by two factors: the triple point of water (the temperature at which water, ice, and water vapour are in equilibrium) is defined to be 0.01 °C, and a degree Celsius equals the same temperature change as a degree on the ideal-gas scale. This puts the boiling point of water at standard atmospheric pressure at 99.975 °C, and not 100° as defined by the Centigrade scale.

Fitting in well with the metric system, the Celsius scale has now largely become the world standard for climatic, industrial, and medical purposes.

Absolute Temperature

Temperature is a measure of the motion – the kinetic energy – of the particles in a material. This raises the question, 'What is the temperature of matter when the molecules have no kinetic energy and are stationary?'

At the beginning of the eighteenth century, when he was investigating the relationship between temperature and pressure in gases, Guillaume Amontons speculated that if the temperature were sufficiently reduced, there would be a point at which there would be no pressure. Pressure is indeed produced by the kinetic energy of the molecules, so he was on the right track.

The Kelvin Scale

In about 1862, the Scottish mathematician and physicist William Thomson, Lord Kelvin (1824–1907), produced the idea of 'absolute zero'. This is a temperature below which it is not possible to go – a temperature at which it is impossible to extract further heat energy from matter. He based his calculations on the laws of thermodynamics and showed that the value of absolute zero is -273.15 °C. The kelvin is now the SI unit of temperature, one of the seven SI base units. This temperature scale is called the absolute, thermodynamic, or kelvin scale, and it is truly an 'absolute' scale because it is the measure of the fundamental property that underlies temperature.

Lord Kelvin
1824–1907

The unit and the scale are defined by two points: absolute zero, and the 'triple point' (see page 83) of pure water. Absolute zero is defined as being precisely 0 K and -273.15 °C, and the triple point of water is defined as being precisely 273.16 K and 0.01 °C. This definition fixes the magnitude of both the kelvin and Celsius units, as well as the relationship between the two scales.

Rankine's Fahrenheit Alternative

A few years later, William J. M. Rankine (1820–1872), a Scottish engineer and scientist, proposed the Kelvin scale in its Fahrenheit form. In this system, absolute zero is 0 °R (degrees Rankine), the value of absolute zero is -459.67 °F, and each Rankine degree is equal to a Fahrenheit degree (whereas each Kelvin is equal to a Celsius degree). This scale is still used in a variety of fields in the U.S. for measuring thermodynamic temperatures, but it is becoming less popular.

The Limits of Coldness

Having identified the theoretical lowest possible temperature, it is reasonable to ask whether this is practically achievable. The answer is, very nearly. For more than 100 years it has been possible to cool matter to below 3 kelvins, but in the last decade or so the frontier has been pushed back even further.

In 1924, building on the work of Satyendra Nath Bose, Albert Einstein predicted that if matter could be cooled to a few hundred-billionths of a degree above absolute zero, the individual atoms would condense together to form a 'superatom' that would behave as a single entity. This 'Holy Grail' of low-temperature physics was labelled the Bose–Einstein condensate, and in 1995 a team working under Eric A. Cornell of the National Institute of Standards and Technology and Carl E. Wieman of the University of Colorado at Boulder finally achieved it.

Rather than slowing down the particles by making them cold, the researchers effectively did the reverse: they made the particles cold by slowing them down. Laser beams fired from all directions were used to batter rubidium atoms into submission, but this 'only' cooled the atoms to about 10-millionths of a degree above absolute zero.

To reduce the temperature further, the atoms were then trapped in a magnetic field. The fastest atoms were allowed to escape and the slowest were retained. In this way, a Bose–Einstein condensate was created. Consisting of some 2,000 rubidium atoms, it formed an entity with a diameter of about one-fifth the thickness of this page.

This strange – and entirely new – form of matter that only occurs close to absolute zero offers quantum scientists a unique opportunity to study subatomic particles in a new way.

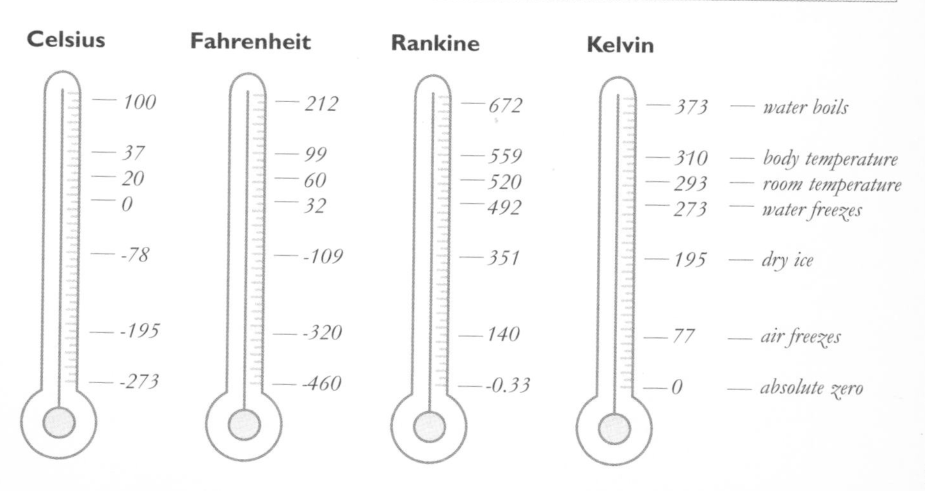

CHAPTER SEVEN

a measure of **time**

Although its definition remains elusive, the accurate measurement of time is central to an understanding of the laws and processes at work in the universe. Since prehistory, humankind has been fascinated by the concept of time and its passage, and its importance to so many aspects of culture now requires it to be coordinated internationally.

'What then is time?
If no-one asks me, I know what it is.
If I wish to explain it to him who asks, I do not know.'
Saint Augustine

Years and Months

The first concepts of time were those based on the apparent movement of the sun and the stars across the sky, the changing appearance of the moon, and the cycles that these reveal.

'Absolute, true, and mathematical time, in and of itself and of its own nature, without reference to anything external, flows uniformly and by another name is called duration. Relative, apparent, and common time is any sensible and external measure (precise or imprecise) of duration by means of motion; such a measure – for example, an hour, a day, a month, a year – is commonly used instead of true time.'

Isaac Newton, Principia

Earliest Measurements of Time

Absolute and true, or relative and apparent, there is evidence, in the form of the notched fragment of a baboon's leg bone found in the Lebombo Mountains between South Africa and Swaziland, that time was being measured by Paleolithic humans 37,000 years ago. The bone is inscribed with 29 lines, corresponding to the number of days in a synodic month – the period between identical phases of the moon, or 'lunations'.

By the third millennium B.C., the Sumerians had a refined calendar based on a wealth of accurate observations and measurements. A good understanding of the annual cycle of the seasons – with all that it means for planting, irrigation, and harvesting – is crucial to a society based upon agriculture, but their calculations and calendar went well beyond the bare necessities. It is from the sexagesimal number system of Mesopotamia (see page 10) that we have inherited our divisions of the hour and the minute. The Sumerians divided the circle into 360 parts, and this was probably the origin of their 360-day year, later amended to 365.

Calendars

The use of the cyclical movements observed in the heavens appears to meet the criteria of the SI metric system, in which units are based on fixed natural phenomena. Unfortunately, however, neither the time taken for the earth to orbit the sun nor the time between the phases of the moon are perfectly

The Lunar Cycle: *As the moon orbits the earth and the earth orbits the sun, the angle between the moon and the sun changes relative to the earth. As a result, our view of the illuminated side of the moon changes in a 29-day cycle.*

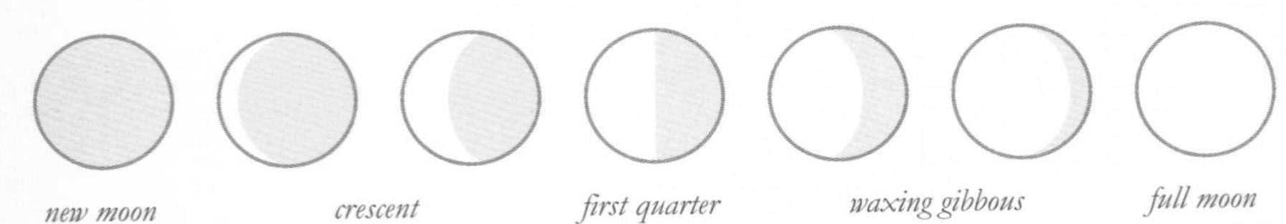

divisible by a whole number of days. Nor is the first divisible by the second, so a lunar calendar and a solar calendar can never be synchronized.

Most civilisations throughout history have adopted a solar calendar of 365 days, adjusting the cycle by adding an extra day every fourth year (leap year) to synchronize the year with the movement of the earth around the sun. In the solar calendar, the months are close to the length of the synodic month, but they are not linked to the phases of the moon.

A lunar year based on 12 lunations will start 11 days earlier each year in relation to the solar year. For this reason, the Hebrew calendar, which is based upon lunations, is actually lunisolar, having lunar months, but solar years that synchronize the religious calendar with the seasons. This is achieved by adding an intercalary (leap) month approximately every three years. The Chinese and Hindu calendars are also lunisolar.

The Islamic calendar is purely lunar, consisting of 12 lunations, and the Islamic religious feasts therefore move 'backward' one full solar cycle in the course of every 33 years.

The date of Easter, the most important feast in the Christian calendar, obeys lunisolar rules and is celebrated on the first Sunday after the full moon that falls on or after the Spring Equinox (March 21).

Multiples of Years

Olympiad: 4 years
In Ancient Greece, an Olympiad was a four-year period from the start of the Olympic Games, which were held at the full moon closest to the summer solstice. The following Olympic Games marked the beginning of the next Olympiad. In modern usage, it can refer to the period leading up to and including the Games.

Lustrum: 5 years
The Romans held a census every five years, after which they made a purificatory sacrifice of animals on behalf of the people. The sacrifice was called a *lustrum*, but the word came to be applied to the census itself and then to a period of five years.

Octaeterid: 8 years
In the Ancient Greek calendar, an *octaeterid* was a period of eight years during which three extra 30-day months were intercalated to bring the lunar year into close accordance with the solar year. The Greek astronomer Meton of Athens later calculated that adding seven months every 19 years gave a closer result.

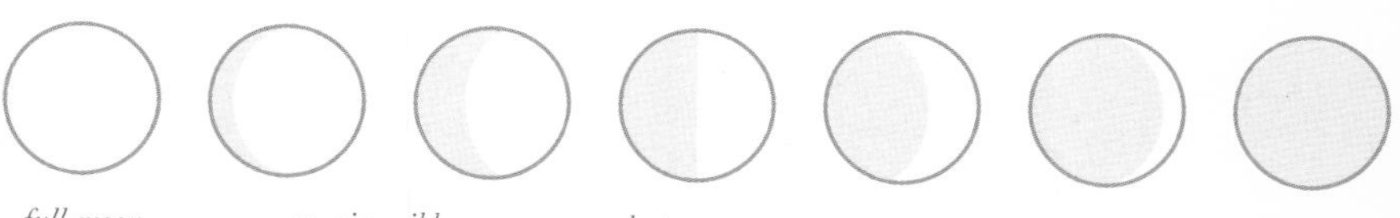

full moon *waning gibbous* *last quarter* *crescent* *new moon*

Time and the Système International

If months and years are not fixed periods of time, what about days and weeks? Could they be used as SI units?

Like years and months, days and weeks are linked to the movement of the earth relative to the sun, so they, too, are variable. Not only is the speed of rotation of the earth irregular, it is also changing. A solar day (which is the day as we know it) is the time taken for a single rotation of the earth relative to the sun, but the precise speed of rotation is gradually decreasing, due to the movement of the world's tides under the influence of the moon's gravity. In fact, the length of the day has increased by a full 45 milliseconds over the last 2,700 years . . . though you may not have noticed it.

SI Time

There was an attempt in revolutionary France to implement a metric day – and the idea still has its proponents – but so far nobody has been tempted to have 10 days in a metric week, 10 metric hours in a day, 100 metric minutes in a metric hour, and 100 metric seconds in a metric minute. (For those who feel that an hour lasting 144 old minutes would be cumbersome, the day could perhaps be divided into centidays of 14.4 old minutes.)

Nonetheless, time being an essential component of the physical world, the SI system does require a base unit of time, and it is the second. Despite all that has been said, the second was at one time defined as 1/86,400 of the mean solar day, leaving the mean solar day to be defined by astronomers. To overcome the problems of the irregularities in the earth's rotation, in 1960 the 11th General Conference on Weights and Measures (CGPM) ratified a definition of the second as 1/31,556,925.9747 of the length of the tropical year 1900. Of course, this posed its own problems, not least that it was rather difficult to go back and check.

A definition based on a natural and fixed phenomenon was finally introduced by the 13th CGPM in 1967–8.

The Right Time

Although SI units do not affect the way we express time, the Time Section of the International Bureau of Weights and Measures is responsible for maintaining International Atomic Time, from which all local time is derived. International Atomic Time is kept to an accuracy of about three milliseconds in 1,000 years, but because the earth's rotation is slightly irregular, we also have Coordinated Universal Time, in which a leap second is periodically added by International Earth Rotation Service in Paris.

The Sidereal Day

This is the amount of time it takes the earth to complete one rotation relative to the stars, and is almost four minutes shorter than a solar day. This is because the earth is orbiting the sun and therefore needs a further four minutes of rotation before the sun appears in the same place in the sky that it was a day earlier.

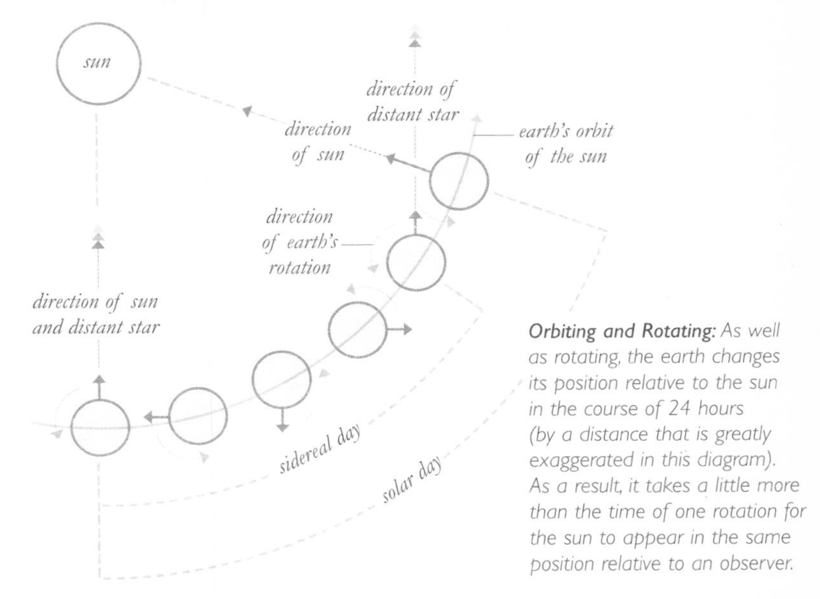

Orbiting and Rotating: *As well as rotating, the earth changes its position relative to the sun in the course of 24 hours (by a distance that is greatly exaggerated in this diagram). As a result, it takes a little more than the time of one rotation for the sun to appear in the same position relative to an observer.*

It had been found that the transition of an atom or molecule between two energy levels was unchanging and could be reproduced accurately anywhere, so the new definition, which still stands, was that the second is 'the duration of 9,192,631,770 periods of the radiation corresponding to the transition between the two hyperfine levels of the ground state of the cesium-133 atom.' In 1997, the International Committee for Weights and Measures added that this definition refers to a cesium atom at rest at a temperature of 0 K (absolute zero).

Prefixes

In the same way that the metre can have prefixes attached, so can the second, but those that multiply the second by powers of ten are rarely used, and the year is not referred to as 31.536 megaseconds. The Système International fortunately has no comment to make on the way in which we express time or dates. These subjects are beyond its scope. The minute, hour, and day – with symbols min, h, and d – are officially considered 'non-SI units accepted for use with the International System.'

Dividing the Second

Prefixes that indicate successive decimal subdivisions of the second are commonly used in scientific contexts. For example, light will travel 1 kilometre in a vacuum in 3.3 microseconds (3.3×10^{-6} s), each step in a nuclear chain reaction takes about 10 nanoseconds (10×10^{-9} s), and rapid chemical reactions take a few hundred femtoseconds (1 fs = 1×10^{-15} s).

In the last ten years, technological advances have made it possible to create and measure laser-light pulses lasting a few femtoseconds, and scientists from the Max Planck Institute for Quantum Optics and Vienna University of Technology have now made a 'stopwatch' that can measure a duration of 100 attoseconds, the time scale on which subatomic events occur. An attosecond is one billionth of a billionth of a second (1 as = 1×10^{-18} s). To put that in perspective, an attosecond is to a second what a minute is to about 20 billion years.

The Smallest Unit of Time

As discussed on page 39, to talk of a length smaller than the Planck length (1.6×10^{-35} metres) is meaningless, because measurement below this scale is logically impossible due to the nature of quantum physics. The shortest possible time duration – the smallest meaningful measurement of time – is linked to this unit. Planck time is the time it would take a photon travelling at the speed of light in a vacuum to cross a distance equal to the Planck length, and it is equivalent to 5.391×10^{-44} seconds. An attosecond may seem small, but it is about 10^{26} (one hundred million billion billion) times bigger than a Planck time.

In theoretical physics, it is not possible to know what happened before 1 Planck time after the Big Bang. It is thought that at this point in time gravitation became differentiated from the other three fundamental 'forces' (electromagnetism, the weak interaction, and the strong interaction). The search for a 'grand unified theory' seeks to explain these three as aspects of a single force.

Max Planck
1850–1938

Measuring Geological Age

We can measure infinitesimal periods of time, but how can we measure time on the grand scale of geological events? The answer lies in the half-lives of radioactive isotopes.

The natural radioactive decay of uranium was discovered by Henri Becquerel in 1896, and Lord Rutherford soon suggested that this phenomenon could be used as a tool for measuring geological age. In 1907, professor Bertram Boltwood, a radiochemist at Yale University, put this idea into practice and demonstrated the enormity of the geological timescale.

Ernest (Lord) Rutherford
1871–1937

Henri Becquerel
1852–1908

Comparing Isotopes

All the atoms of an element have the same number of protons in their nuclei, but the number of neutrons can differ, giving the atoms different atomic weights. Atoms of the same element with differing atomic weights are called isotopes. Some isotopes undergo radioactive decay, losing particles from their nuclei to become an isotope of a different element. The rate at which this spontaneous process occurs varies between radioactive isotopes, and can be expressed as the 'half-life' of that isotope – the period of time taken for half the atoms in a 'parent' sample to decay and become the 'daughter' isotope. Depending on the isotope under consideration, the half-life can be anything from a matter of minutes to billions of years.

Isotopes that decay slowly to produce stable daughter isotopes (i.e. isotopes that do not decay further) can be used as geological clocks. This is because in a sample of rock the relative proportions of the parent and daughter isotopes will reveal what proportion of the parent isotope's half-life has passed since the rock was created.

Some commonly used isotopes include:

Parent Isotope	*Daughter Isotope*	*Half-Life (years)*
Uranium-235	Lead-207	704 million
Potassium-40	Argon-40	1.25 billion
Uranium-238	Lead-206	4.5 billion
Thorium-232	Lead-208	14.0 billion
Rubidium-87	Strontium-87	48.8 billion
Samarium-147	Neodymium-143	106 billion

Carbon Dating

For shorter periods of time, the radioactive isotope carbon-14 – known as radiocarbon – offers a useful means of measurement. Cosmic radiation creates carbon-14 (which has a half-life of 5,730 years) continuously in the upper atmosphere, and this is taken up by all living plants and animals, along with non-radioactive carbon. All living things therefore contain the same proportions of radiocarbon and non-radioactive carbon – approximately one part to one trillion. When an organism dies, the amount of carbon-14 in it starts to decline as it decays to become nitrogen-14, so the amount of radioactive carbon that remains indicates how long ago the organism died. If, for example, a bone contains only half as much carbon-14 as bone from a living creature, that bone is 5,730 years old. Carbon-14 dating, which can only be used on organic matter, is most useful for dating material from the last 50,000 years.

However, there are some questions about the accuracy of the method when used on samples more than a few thousand years old. The model assumes that:

- radioactive decay has always taken place at the same rate
- the level of cosmic radiation (and therefore the rate of production of carbon-14) has been constant
- the concentration of non-radioactive carbon in the atmosphere does not fluctuate
- the ratio of radiocarbon to non-radioactive carbon in the atmosphere is therefore constant.

How Old Is the Earth?

Radiometric methods date the oldest rocks on earth at about four billion years, but the dating of meteorites shows them all to be about 4.56 billion years old. Meteorites are pieces of asteroid. These small bodies were formed in space and probably cooled fairly quickly, unlike the earth, so this is likely to be a realistic measure of their true age. If, as most scientists believe, all the bodies in the solar system came into existence at about the same time, this is also the age of the earth.

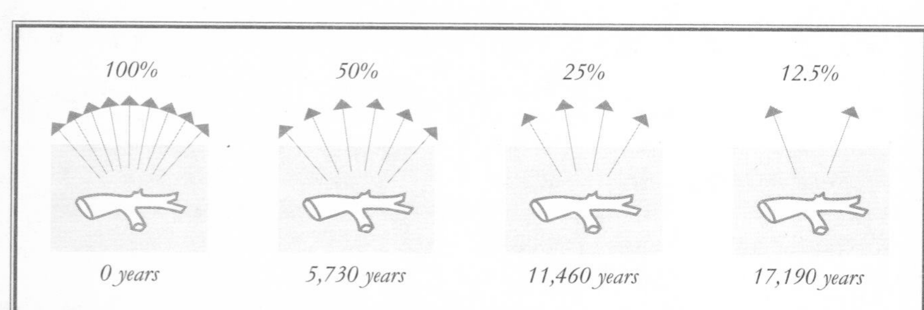

Carbon Decay: *There is a fixed proportion of carbon-14 in living things, but once it dies this proportion halves every 5,730 years. The rate of decay in a sample of carbon of known weight reveals how long ago the organism died.*

Geological Timeline

The earth's timescale is divided into time periods corresponding to observable geological strata. Much of the terminology is based on research carried out in Europe in the nineteenth century. The fine detail of the timescale is continually being refined, and not all subdivisions have been shown here.

MYA = million years ago

Eon	Era	Period	Epoch
Phanerozoic (543 MYA to present)	Cenozoic (65 MYA to today)	Quaternary (1.8 MYA to today)	Holocene (10,000 years to today) Pleistocene (1.8 MYA to 10,000 yrs)
		Tertiary (65 to 1.8 MYA)	Pliocene (5.3 to 1.8 MYA) Miocene (23.8 to 5.3 MYA) Oligocene (33.7 to 23.8 MYA) Eocene (54.8 to 33.7 MYA) Paleocene (65 to 54.8 MYA)
	Mesozoic (248 to 65 MYA)	Cretaceous (144 to 65 MYA) Jurassic (206 to 144 MYA) Triassic (248 to 206 MYA)	
	Paleozoic (543 to 248 MYA)	Permian (290 to 248 MYA)	
		Carboniferous (354 to 290 MYA)	Pennsylvanian (323 to 290 MYA) Mississippian (354 to 323 MYA)
		Devonian (417 to 354 MYA) Silurian (443 to 417 MYA) Ordovician (490 to 443 MYA)	
		Cambrian (543 to 490 MYA)	Tommotian (530 to 527 MYA)
Proterozoic (2,500 to 543 MYA)	Eras: Neoproterozoic (1,000 to 543 MYA) / Mesoproterozoic (1,600 to 1,000 MYA) / Paleoproterozoic (2,500 to 1,600 MYA)		
Archaean (3,800 to 2,500 MYA)	Eras: Eoarchean (3,800 to 3,600 MYA) / Paleoarchean (3,600 to 3,200 MYA) / Mesoarchean (3,200 to 2,800 MYA) / Neoarchean (2,800 to 2,500 MYA)		
Hadean (4,500 to 3,800 MYA)			

CHAPTER EIGHT

a measure of **speed**

Speed is a measure of the rate of movement of an object – the distance that it travels in a given period of time. Units of distance and of time have, of course, already been explained. Speed, however, is much more than the product of its two parts.

'It is impossible to travel faster than the speed of light, and certainly not desirable, as one's hat keeps blowing off.'
Woody Allen, Film Director

SI Units of Speed

The base unit of distance in the SI system is, of course, the metre, and the base unit of time is the second, so the derived SI unit of speed is metres per second (m/s).

The SI units of speed contain the first surprise, or at least an apparent contradiction. The metre, as we have seen, is defined as the distance that light travels in an absolute vacuum in 1/299,792,458 seconds. The speed of light (see page 102) is 299,792,458 metres per second. This might lead to the assumption that there needs to be a definition of the metre other than one based on the speed of light. After all, there is surely a contradiction in defining two things in terms of each other?

In fact, the length of the metre was originally defined as 1/10,000,000 of the distance from the Equator to the North Pole along the meridian that passes through Paris (see page 34). Once it was realised that the speed of light is a constant, it then made sense to define the metre in terms of that speed and in terms of the second, which is now also defined by reference to a natural and invariable phenomenon (see page 91). Contradiction resolved.

Ease of Use

As with all the systems of units that we have looked at, it is essential to create units that are practical and meaningful. If you are driving from Richmond to Rochester and you only know your speed in metres per second, working out how long the journey is likely to take will give you a headache. (It's easy enough to go from metres to kilometres just by moving the decimal point, but the units of time are not based on a decimal system.) Countries that use the SI system therefore express the speed of a car in kilometres per hour (km/h), which provides a much more usable answer to the problem, although the hour is not an SI unit.

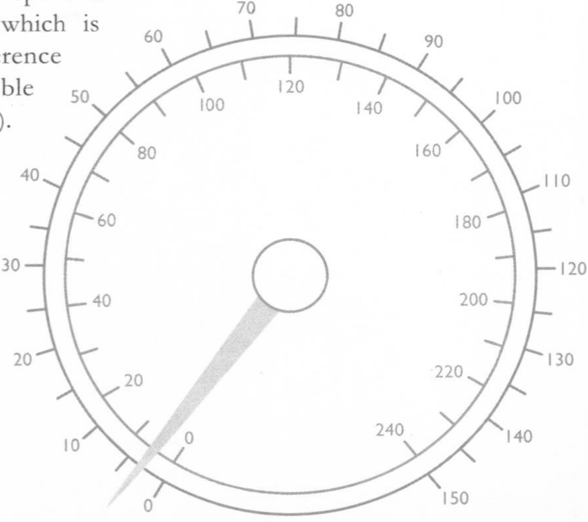

***Speed:** Speedometers on vehicles destined for the U.S. and the U.K. display mph as the primary scale but usually also show km/h. Most other countries require only km/h. In this diagram, the inner ring shows km/h and the outer mph.*

Average Speed

In many cases, the speed of a moving object varies. The reading on the speedometer of a car, for example, will change continuously. The car does not move the same distance in every unit of time, so the 'instantaneous speed' – the speed at any given moment – is changing all the time. The average speed of an object is often, therefore, more useful.

Average speed is the total distance moved divided by the total time. If a bicycle has been ridden for an hour, and covered 18 km during that time, the average speed is therefore 18 km/h (or 5 m/s), even though it might have been travelling at this exact speed for only a small fraction of the journey.

Speed or Velocity?

Although the terms are often used interchangeably, and despite the fact that they can be expressed in the same units, speed and velocity are not the same thing. The time element is the same in both cases, but the concept of distance is different for each. In the case of speed, the distance is a 'scalar' quantity, which means that it only has one dimension – that of magnitude. It is the distance that a body moves 'over the ground', regardless of direction. Distance in the case of velocity should really be referred to as 'displacement' – the change in position of the body. This is a 'vector' quantity, having the dimensions of both magnitude and direction.

Returning to the example of a bicycle being ridden for an hour and covering 18 km, let us suppose that the rider were cycling along winding country roads and that at the end of the hour they were actually 10 km northwest of their starting point. We have already seen that their average speed was 18 km/h, but their average velocity was 10 km/h northwest.

If we suppose that the journey started and ended in the same place, then the average velocity would be zero (average velocity = total displacement divided by total time, and the displacement is zero).

Just to confuse matters, the symbol for speed is v, and the equation for speed is given as $v = d/t$, where d is the distance and t is the time.

Distance vs. Displacement: *The dotted gray line shows the actual path – the distance – travelled by a cyclist. The straight line shows the displacement.*

the length of this line divided by the time taken gives the average speed

the length of this line (and its direction) divided by the time taken gives the velocity

Imperial and U.S. Customary Units of Speed

As we have seen, the old English system of units contained a plethora of units of length, and many of these have historically been divided by various units of time to create a range of ways of expressing speed.

The Imperial system whittled the units of length down to a reasonable number, and the U.S. Customary system retains only the inch, foot, yard, and mile. There is no accepted 'base unit' of speed, and any of these units of length (although rarely the yard) can be used to create units of speed that are appropriate to the nature of the phenomenon or object concerned.

MPH AND FPS

In the U.S., where the unit of length on maps and road signs is the mile, the primary scale on the speedometer of a vehicle shows miles per hour (written as mph or MPH). Often, a secondary scale also indicates the speed in kilometres per hour (km/h), in recognition of the gradual adoption of metric measurements, and presumably to help

Wind Speeds

The speed and direction of the wind are critical factors in determining weather patterns. They are also of immediate interest to shipping, and weather maps therefore need to display this information in a clear and quickly understood way. Meteorological data is collected continuously from a variety of weather 'platforms', including buoys, weather ships, coastal weather stations, and airports, and the wind conditions are then shown on weather maps using 'wind barbs'. The circle at the end of the long line represents the weather platform, and is positioned over it on the map.

The direction in which the long line points represents the direction from which the wind is blowing (in the examples shown here, from the east).

The wind speed is shown by the shorter lines, or barbs. Each long barb represents 10 knots; each short barb represents 5 knots. A triangular pennant represents 50 knots.

calm

5 knots

10 knots

15 knots

20 knots

50 knots

65 knots

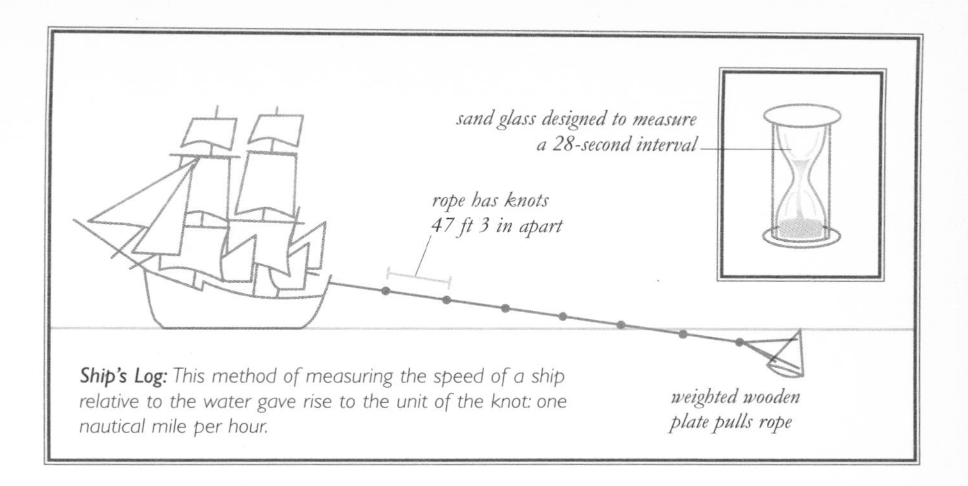

Ship's Log: *This method of measuring the speed of a ship relative to the water gave rise to the unit of the knot: one nautical mile per hour.*

drivers avoid breaking the law when they cross into Canada, where kilometres are used. A small minority of road signs in the U.S. also show distances in kilometres (a very small number show just the metric measurement), but speed restriction signs on highways generally display limits only in miles per hour.

Although the SI derived units of speed are used in the U.S. in scientific and medical fields, feet per second (ft/s, ft/sec, or fps) are still used in the military when referring to the muzzle velocity of a firearm, and the speed of ships and aircraft is commonly given in knots (see below).

Perhaps surprisingly, in the U.K., which purports to have completed the switch to the metric system, road signs and the principal scale on speedometers and odometers still give speeds and distances in miles and miles per hour respectively. (Since gasoline is now sold by the litre, however, this complicates the task of working out the fuel economy of a car in either miles per gallon or litres per 100 kilometres.)

Nautical miles per hour

As we have seen, the nautical mile is the equivalent of a minute of arc on a great circle (see page 33). The derived unit of speed, defined as one nautical mile per hour, is the knot, and although it has absolutely no relationship with the SI system, it is in everyday use for navigation by sea and air.

The term 'knot' comes from the naval practice of using a knotted rope to gauge a ship's speed. On sailing ships, a weighted wooden drogue would be thrown overboard, attached to a rope with a knot tied in it every 47 feet 3 inches. As the ship moved through the water, a sailor counted the number of knots that passed through his fingers in a 28-second period, the time being measured with a sand glass. The number of 'knots' was the speed of the ship in nautical miles per hour.

The distance between the knots is the key. There are 3,600 seconds in an hour. Divide this by 28 and multiply by 47.25 and you get 6,075, a very close approximation to the Admiralty nautical mile of 6,080 feet.

The Speed of Light and Sound

As we have seen in several contexts, the speed of light (and of all other electromagnetic waves) in a vacuum is constant. This is so not just because the definition of the metre is linked to the speed of light. The two statements 'A metre is the distance travelled by light in a vacuum in 1/299,792,458 seconds,' and 'The speed of light is 299,792,458 metres per second' do indeed mean that the speed of light is a constant by definition, but it has also been proven to be so both theoretically and experimentally, and using different definitions of the metre.

The speed of light plays an important role as a constant (denoted by the symbol c) in quantum physics, and most famously in Einstein's equation $E = mc^2$ (energy = mass x the speed of light squared), which states the relationship between energy and mass and was vital in the development of the atomic bomb.

A Leading Question

When discussing the constancy of the speed of light, a question that has an unexpectedly interesting answer is: 'The speed of light relative to what?' When we talk about the speed of sound we generally mean the speed at which sound waves move relative to the medium through which they are passing, but light doesn't need a medium through which to pass. Another logical answer might be 'relative to the source of the light,' rather like the speed of a bullet relative to the rifle that fired it, but it has been found that the speed of light from a source that is moving away is the same as that from a source that is moving toward the observer.

It can also be proven mathematically that if two observers are moving relative to a source of light, one of them away from the source at half the speed of light and the other toward it at the same speed, they will both find that the light is travelling at 299,792,458 metres per second. This certainly defies common sense, but it is an unavoidable consequence of Einstein's Theory of Relativity. The answer to the initial question is 'The speed of light relative to the observer.'

Ernst Mach
1838–1916

Mach I

Unlike the speed of light, the speed of sound is not a universal constant. Sound can only travel through an elastic medium (not a vacuum), and the properties of that medium – especially its temperature in the case of a gas – affect the speed at which sound waves are propagated through it. Nonetheless, because of its significance for aircraft, the speed of sound is used as the base unit in the Mach scale (named after the nineteenth-century Austrian physicist Ernst Mach).

What's the Big Noise?

The speed of sound is important for aircraft because they experience certain aerodynamic effects when travelling at this speed. One of these effects produces the familiar 'sonic boom' that occurs when an airplane exceeds the speed of sound. As it moves through the air, a plane compresses the air in front of it, forming waves that travel out from the craft at the speed of sound. As the airplane approaches the speed of sound, it effectively catches up with the waves that are emanating in front of it and the compressions build up into a single pressure wave at the speed of sound. A similar wave forms behind the plane as the air pressure returns to normal, and these shock waves are heard on the ground as a double boom. This boom is generated continuously while the aircraft is supersonic, and the area below the flight path along which it can be heard is called the 'boom carpet'.

The Mach Number

The speed of sound in air at sea level at a temperature of 59 °F (15 °C) and under normal atmospheric conditions is 761 mph (1225 km/h or 340 m/s), but at high altitudes, where the temperature can be significantly lower, sound travels more slowly. An airplane is said to be travelling at Mach 1 when it reaches the speed of sound under those specific conditions. Mach 2 is twice as fast. The Mach number is the ratio of an object's speed to the speed of sound in the medium through which it is passing.

Sonic Boom

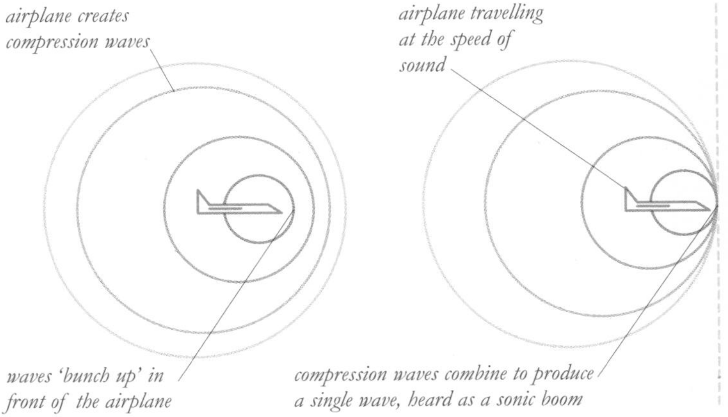

Wall of Sound: *At the speed of sound, an airplane catches up with the compression waves it is creating, producing a single shock wave.*

The Quest for c

The speed of light (c) – and, indeed, of all electromagnetic radiation – has proved to be a vitally important constant, but how does one measure something travelling at about 186,000 miles (300,000 kilometres) per second?

Galileo

At a time when the prevailing view was that the speed of light was infinite, Galileo (1564–1642) tried to prove it was finite and to measure it by placing two people some distance apart, each with a covered lantern. The first flashed his lantern, the second did the same when he saw the light, and the first attempted to detect how much the delay changed with distance. A distance of a few miles was insufficient to produce a measurement.

Roemer

The first person to come up with a good approximation was the Danish astronomer Ole Roemer (1644–1710), working at the Paris Observatory in 1675. Attempting to explain why the period of orbit of Io, one of the moons of Jupiter, changed with the time of year, he realised that the light from Jupiter took longer to reach the earth when the two planets were on opposite sides of the sun, causing an apparent delay in Io's reappearance from behind Jupiter. From the delay and the calculated distance, he found the speed of light to be 140,000 miles (225,000 km) per second, an error of only 25 percent.

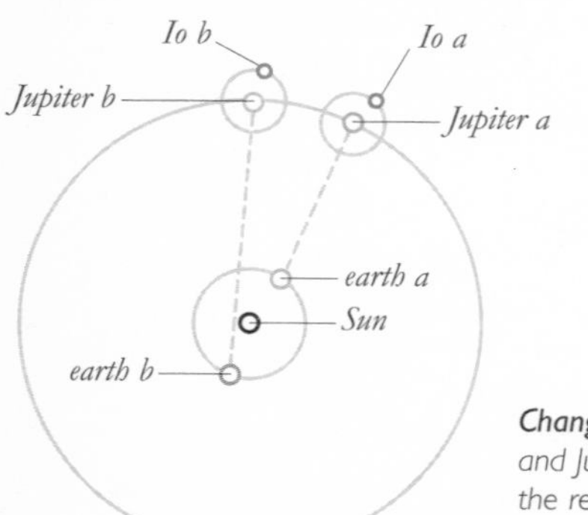

Changing Distance: *As the distance between the earth and Jupiter increases, the time taken for the light from the reappearing moon Io to reach earth increases, delaying Io's reappearance.*

Bradley

In 1728, the English astronomer James Bradley (1693–1762) calculated the speed of light from the apparent displacement of stars due to the motion of the earth around the sun, on the basis that the direction in which the observer is travelling affects the apparent angle of the incident light. His value for the speed of light was 187,000 miles (301,000 km) per second.

Fizeau

The French physicist Armand Fizeau (1819–1896) made the first attempt since Galileo to measure the speed of light in air. In 1849, he set up an apparatus consisting of a light source, two mirrors, and a rotating wheel with slits in it. The light was reflected from the first mirror through a slit in the rotating

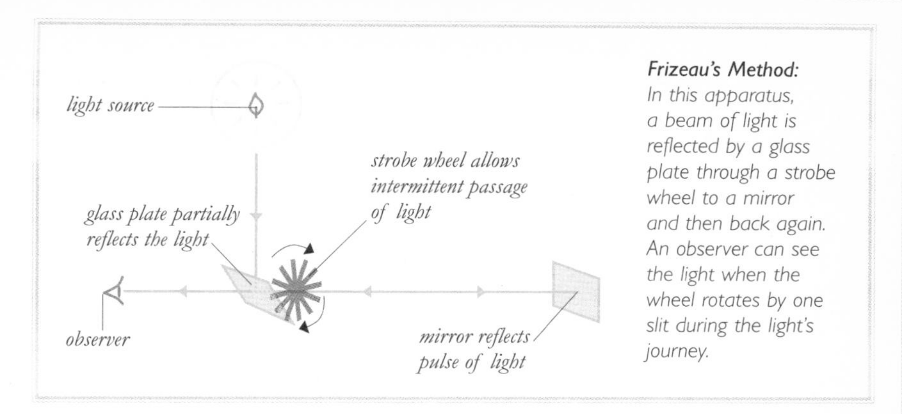

Frizeau's Method: *In this apparatus, a beam of light is reflected by a glass plate through a strobe wheel to a mirror and then back again. An observer can see the light when the wheel rotates by one slit during the light's journey.*

wheel, creating a short pulse of light that travelled 5 miles (8 km), to be reflected from the second mirror back through the slotted wheel to the observer. when the wheel was rotating slowly the light returned through the same slit, but as the wheel was speeded up the pulse hit a blade of the wheel. At a certain speed the light could be seen through the next slit, and by knowing the speed of rotation Fizeau was able to calculate the time taken for the light to make the 10-mile (16-km) round trip, and hence the speed of light. He came up with the figure of 195,000 miles (315,000 km) per second.

Foucault

Using a method invented by British scientist Sir Charles Wheatstone (1802–1875) to measure the speed at which electricity travels through a wire, using the speed of a rotating mirror to determine the delay between two events, the French physicist Léon Foucault (1819–1868) improved on Fizeau's apparatus a year later. At a certain speed, the revolving mirror deflected the returning beam of light through a measurable angle, which enabled Foucault to know how far the mirror had turned and, from the mirror's speed of rotation, to work out the time taken by the light. His figure of 185,000 miles (298,000 km) per second was the most accurate yet.

Twentieth-Century Refinements

American astronomer Simon Newcomb (1835–1909) and optical scientist Albert Michelson (1852–1931), the first American to win the Nobel Prize in physics, refined the method further between 1879 and 1927, producing ever more accurate figures. In this same period, physicists basing their work on Maxwell's theory of electromagnetism were also increasing the accuracy with which the speed of light was known. In 1907, E. B. Rosa and N. E. Dorsey at the National Bureau of Standards obtained a figure of 299,788 km/s using an electrical method.

In the second half of the twentieth century, the development of microwave interferometry, lasers, and cesium clocks steadily improved measuring techniques, and by 1973 the speed of light was known to be 299,792.4574 km/s. The adopted value is 299,792,458 m/s.

Acceleration

VELOCITY IS THE RATE OF CHANGE OF POSITION WITH RESPECT TO TIME. THE RATE OF CHANGE OF VELOCITY WITH RESPECT TO TIME IS TERMED ACCELERATION. WHILE VELOCITY IS EXPRESSED IN UNITS OF LENGTH PER UNIT TIME, ACCELERATION IS EXPRESSED IN UNITS OF VELOCITY PER UNIT TIME, OR LENGTH PER UNIT TIME PER UNIT TIME.

Take the example of a car accelerating uniformly from a standing start. It is travelling at 10 feet per second (about 3 m/s) after one second, 20 fps (6 m/s) after two, 30 fps (9 m/s) after three, and so on. The acceleration of the car is expressed as 10 feet per second per second or 10 fps^2 ($3\ m/s^2$). Since the velocity increases by the same amount in each unit of time, it is said to have constant acceleration.

The units of time used in the expression of the velocity need not be the same as those used to express the change in velocity, so one could express the acceleration of a car in terms of the change of speed in miles per hour per second.

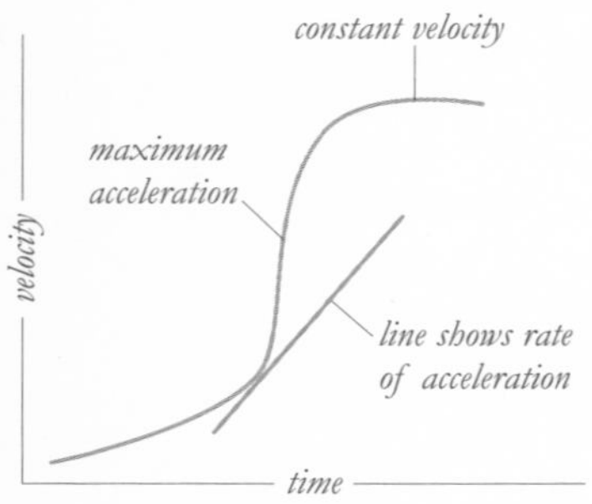

Rate of Acceleration: *When the velocity of a moving object is plotted against time, the angle of the curve at any point (shown by a tangential line) shows the rate at which the velocity is changing – in other words, the rate of acceleration.*

We have already seen that velocity is a vector quantity, meaning that as well as speed it has the dimension of direction. Since acceleration expresses the rate of change of velocity, it too has the dimension of direction. If the velocity is in a straight line in the direction to which we assign a positive value and the velocity is increasing, then the acceleration is positive. If the velocity is decreasing, the acceleration is negative. Conversely, if the direction of movement is negative and the velocity is decreasing, this is a positive acceleration, and if the velocity is increasing, the acceleration is negative.

Gravity

An object in free fall under the influence of gravity accelerates at a constant rate. The actual acceleration varies around the globe, but the symbol g is used for the average acceleration produced at sea level. By international agreement, the standard acceleration due to gravity g_n is about 32.17405 feet per second per second – it is actually defined as exactly 9.80665 metres per second per second (m/s^2).

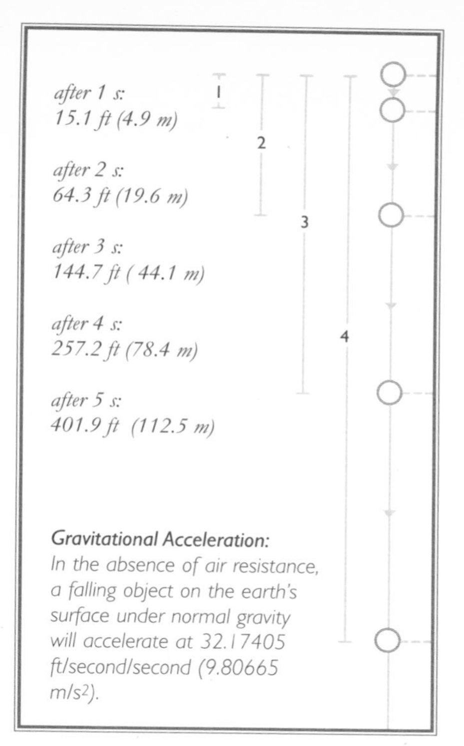

Gravitational Acceleration: *In the absence of air resistance, a falling object on the earth's surface under normal gravity will accelerate at 32.17405 ft/second/second (9.80665 m/s^2).*

Angular Acceleration

When an object is travelling in a circle, it has a speed and a direction of rotation. Together these constitute the angular velocity, and this is expressed in radians per second (rad/s) in the SI system (see below). Anticlockwise rotation is described as positive, while clockwise rotation is negative. Other non-SI units such as revolutions per minute (rpm) are also often used.

If the angular velocity of the object is not constant, then the rate of change of angular velocity is called angular acceleration and, with the same logic as the relationship between velocity and acceleration, angular acceleration is expresses in radians per second per second (rad/s^2) in the SI system.

The Radian

A radian is defined as the angle subtended at the centre of a circle by an arc along the circumference that is the same length as the radius of the circle. (Since the circumference of a circle is 2 pi x the radius, there are 2 pi radians in 360 degrees.)

In theory, the full range of SI prefixes can be applied to the radian, but in practice any unit larger than the radian (e.g. decaradian) will describe more than one full circle, so these are not generally used. Smaller units, such as the milliradian, do have their uses. For example, if the projectile from a gun aimed at a target 100 metres away misses the target by one metre horizontally, the gun needs to be rotated .01 radian, or one centiradian, to correct it.

A similar unit of angle measure, called the mil, was used for this purpose by the U.S. Army. Its definition has changed over time, and in the context of target shooting the mil is now used to mean one milliradian, or 0.001 radian, which equates to a horizontal displacement of one centimetre at a range of 10 metres (or .36 inches at 10 yards).

Radian: *The angle subtended by an arc on the circumference of a circle whose length is equal to the radius of the circle.*

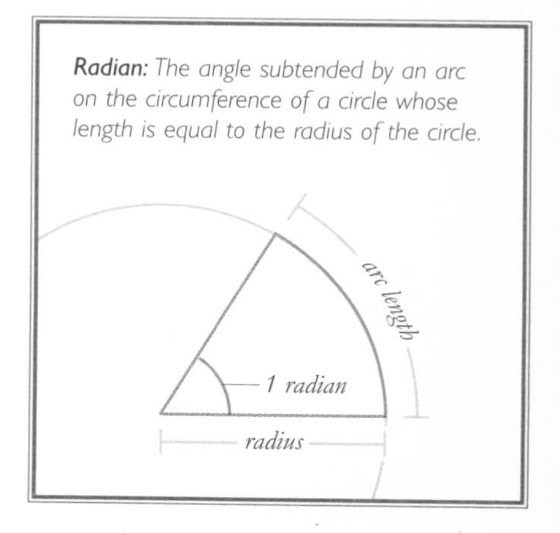

CHAPTER NINE

a measure of force and pressure

On everything around us, from orbiting planets and spinning electrons to cars on the freeway and wind in the trees, force and pressure are in action. Many of the greatest scientists have contributed to this field of physics, enabling us to understand the principles that shape the universe. The units that we use to measure and describe these phenomena give us an insight into the processes at work.

'It is a mathematical fact that the casting of this pebble from my hand alters the centre of gravity of the universe.'

Thomas Carlyle, Scottish Essayist (1795–1881)

What Is Force?

The word 'force' has a host of meanings in everyday language, from the force of an argument to the forces of nature. However, when it is used in the context of physics, the meaning is quite specific.

Force is 'a vector quantity that tends to produce an acceleration of a body in the direction of its application.' It is that which exerts a pull or a push on an object. Since it is a vector quantity, it has the dimensions of both magnitude and direction.

Inertial Mass

At the end of the sixteenth century, the Italian scientist Galileo Galilei formulated of the concept of inertia: an object in a state of motion possesses an 'inertia' that causes it to remain in that state of motion unless an external force acts on it. Inertia is the natural tendency of an object to resist a change in its state of motion. (Under normal circumstances, the forces of friction and air resistance cause moving objects to slow down and stop, but in the absence of any forces they continue to move at a constant velocity.)

Building on the work of Galileo, Sir Isaac Newton was the first to describe mathematically the interaction between forces and objects, and he formulated his Three Laws of Motion. These state:

1. Every object in a state of uniform motion (including a stationary object) tends to remain in that state of motion unless an external force is applied to it.

2. The relationship between the mass of an object (m), its acceleration (a), and the applied force (F) is F = ma (Force = mass x acceleration). In other words, the acceleration of an object is directly proportional to the magnitude of the force acting on it, is in the same direction as that force, and is inversely proportional to the mass of the object. The greater the force, the greater the acceleration of a given mass. The greater the mass, the smaller the acceleration produced by a given force.

3. For every action there is an equal and opposite reaction. If you push against a wall, the wall will exert an equal and opposite force against you. If you and a friend (who has the same mass as you)

Galileo Galilei

The Italian scientist and philosopher Galileo Galilei (1564–1642) studied music, mathematics, geometry, and astronomy. Through meticulous observation and measurement, he formulated laws governing falling bodies and pendulums, and confirmed Copernicus' view that the earth was not the centre of the universe and that the planets revolved around the sun.

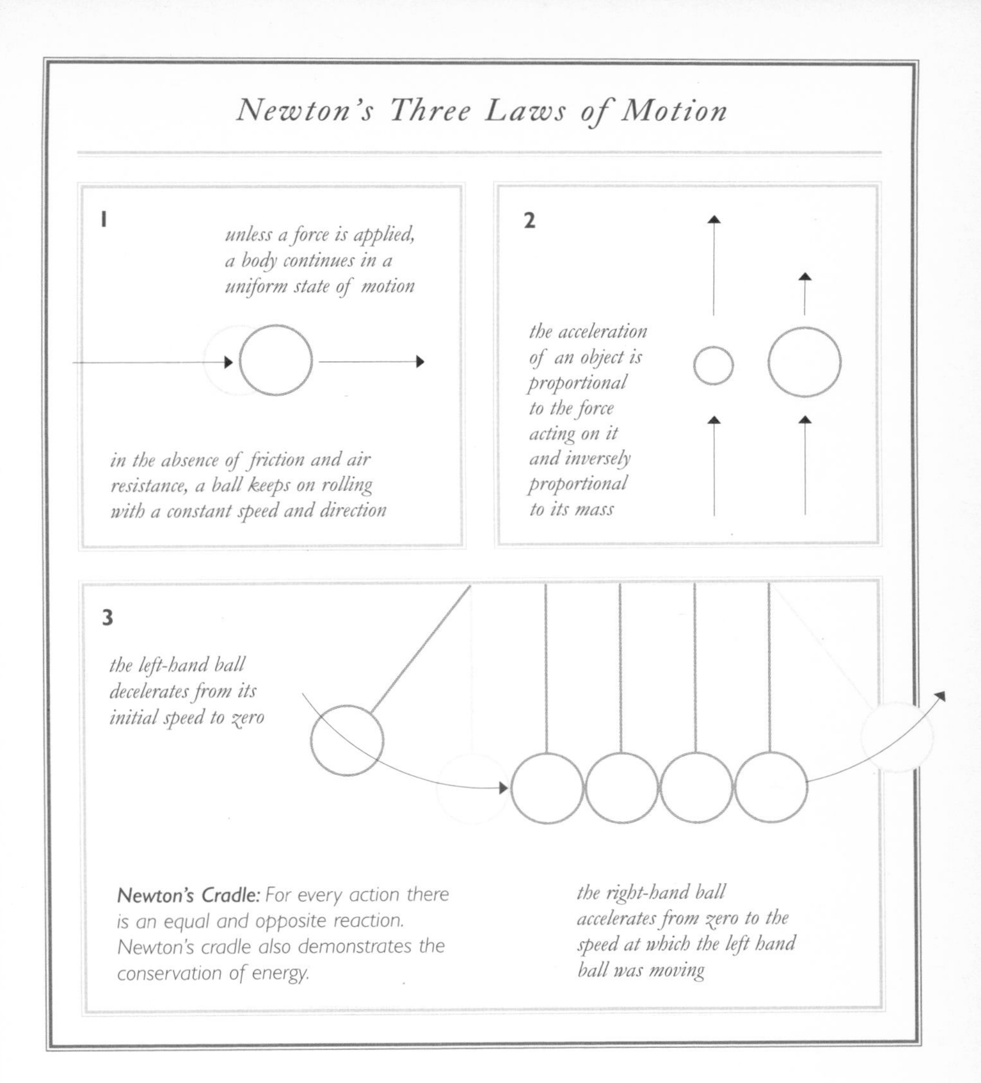

are standing on ice and you push your friend, you will both experience the same acceleration in opposite directions.

Important Implications

Even when we take on board the findings of Galileo and Newton, we still tend to think that a force produces motion. Newton's first law tells us categorically that it does not. A force produces acceleration – a change of motion – and that can mean an increase or decrease in speed and/or a change of direction. If no forces are acting on an object it will remain in the same state of motion, be it stationary or moving with constant velocity – i.e. in the same direction at the same speed. No force is required for an object to remain in motion.

SI and U.S. Customary Units of Force

Force is a complex area of physics, and units of force introduce an equally complex world of newtons, dynes, poundals, and slugs.

The SI Unit of Force

Newton's second law tells us that a force can be defined in terms of the acceleration that it will produce in an object of a given mass (F = ma). In the SI system, the unit of force is, appropriately, the newton (N). The newton is that force which, when acting for one second on one kilogram of matter, will increase its speed by one metre per second.

This is a derived unit, using the kilogram, the metre, and the second. If, in the equation F = ma, mass is expressed in kilograms (kg) and acceleration is expressed in metres per second per second ($m \cdot s^{-2}$), the resulting value for force will be in newtons. 1 N = 1 kg metre per second per second (1 $kg \cdot m \cdot s^{-2}$)

Earlier Decimal Units of Force

In the centimetre–gram–second (CGS) system of units that predates the SI system, the unit of force is the 'dyne'. The dyne is that force which, when acting for one second on one gram of matter, will increase its speed by one centimetre per second. Comparing this definition with that of the newton (above) it can be seen that 1 newton = exactly 10^5 (100,000) dynes.

Another pre-SI unit, and one that is now frowned upon, is the kilogram-force. This is the mass of a kilogram multiplied by the standard acceleration due to gravity. In other words, it is the force exerted on a mass of one kilogram on the surface of the earth. The gravitational pull at the earth's surface actually varies from the poles to the equator, so a standard value of exactly 9.80665 m/s^2 is used. One kilogram-force is therefore, by definition, equal to 9.80665 newtons.

U.S. Customary Units of Force

Although, as we have seen (page 68), the pound is a unit of mass in the foot–pound–second system, it is frequently used as a unit of force. In this context it should be referred to as a pound-force if there is any possibility of ambiguity. The pound-force is defined as the mass of a pound multiplied by the standard acceleration due to gravity (see opposite). In other words, it is the force exerted on a mass of one pound on the surface of Earth.

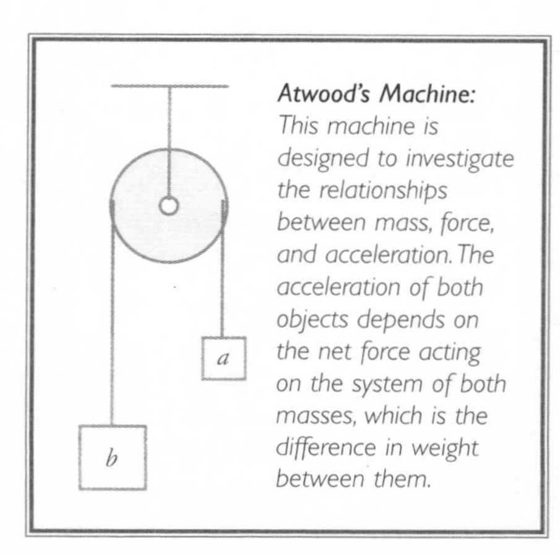

Atwood's Machine: *This machine is designed to investigate the relationships between mass, force, and acceleration. The acceleration of both objects depends on the net force acting on the system of both masses, which is the difference in weight between them.*

Sir Isaac Newton

Sir Isaac Newton
1643–1727

A measure of Sir Isaac Newton's importance to the scientific endeavour is the range of the areas of physics to which he contributed. Probably best known for formulating the concept of gravity, he made major contributions to astronomy, classical mechanics, optics, chemistry, mathematics, thermodynamics, and to our understanding of angular momentum and the propagation of sound. In short, he was a proponent of 'natural philosophy' (the objective study of nature and the universe). He was also an alchemist and a strongly religious man who wrote extensively from a Christian perspective.

One disadvantage of this unit is that, in the equation F = ma, the numerical value of the standard acceleration due to gravity (approximately 32.2 feet per second per second) must be reintroduced as a constant on the right-hand side of the equation in order to achieve a force expressed in pounds-force.

This problem is avoided by using the 'poundal' as the unit of force instead. This is the force necessary to accelerate a pound of mass at one foot per second per second (1 lb·ft·s^{-2}). A poundal is therefore $^{1}/_{32.2}$ pound-force, obviating the need to introduce the constant.

The Slug

The 'slug' (see page 69) gets around the problem by another means. The slug is that mass which, when a force of one pound-force is exerted on it, accelerates by one foot per second per second, and has a mass of approximately 32.2 pounds. Instead of reducing the value of the acceleration by a factor of 32.2 as the poundal does, it achieves the same end by multiplying the mass.

Equivalences Between Units of Force

Newton: = 1 N
= 1 kg·m/s^2
= 105 dyn
= 0.22481 pound-force
= 7.2330 poundal

Dyne: = 1 dyn
= 1 g·cm/s^2
= 10^{-5} N
= 2.2481×10^{-6} pound-force
= 7.2330×10^{-5} poundal

Pound-force: = 1 lbf
= gn*·(1 lb)
= 4.448222 N
= 444822 dyn
= 32.174 poundal

Poundal: = 1 pdl
= 1 lb·ft/s^2
= 0.138255 N
= 13825 dyn
= 0.031081 pound-force

** gn is the standard acceleration due to gravity*

Gravity and Weight

Forces result from interactions between objects. A force may be a contact force (e.g. friction or air resistance), or it may be action-at-a-distance (e.g. gravitational or magnetic force).

According to Newton's assistant at the royal mint, John Conduitt, 'In the year 1666 . . . whilst he was musing in a garden it came into his thought that the same power of gravity (which made an apple fall from the tree to the ground) was not limited to a certain distance from the earth but must extend much farther than was usually thought – Why not as high as the Moon said he to himself and if so that must influence her motion and perhaps retain her in her orbit, whereupon he fell a calculating what would be the effect of that supposition . . .'

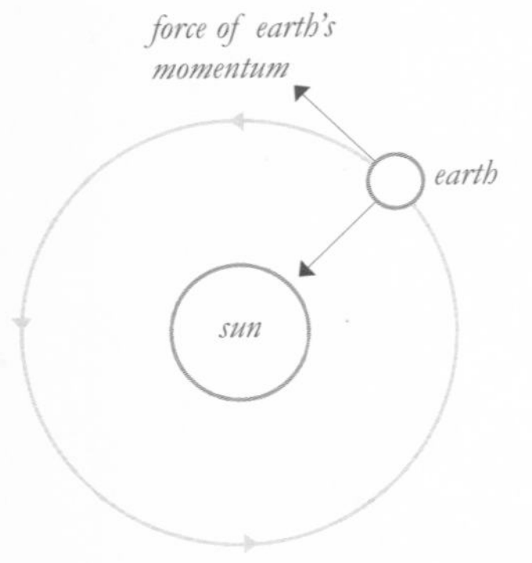

Balancing Forces: *Why doesn't the earth fall toward the sun? The answer is, it does, but at a rate that keeps it on a circular path. The sun's gravity balances the earth's momentum and prevents the planet continuing in a straight line.*

Newton was subsequently able to demonstrate that every object that has mass attracts every other object that has mass, and that the magnitude of the attraction depends on the magnitude of the masses and the distance between them. It is this attraction – gravity – that keeps the planets in their orbits around the Sun, that keeps the moon in its orbit around the earth, and that causes the apple to fall.

Newton also discovered that the force of attraction between two objects is proportional to the masses of the two objects (m_1 and m_2), and is inversely proportional to the square of the distance between them (r). That is, F is proportional to $(m^1 \times m^2)/r^2$. If the mass of either is doubled, the force doubles. If the distance between the objects is doubled, the force is multiplied by $1/2^2$ (i.e. it is a quarter as great).

Force and Weight

We experience this force – the force of gravity – as an upward force on our feet when we are standing, and as the upward force of the chair when we are sitting. We even have an easy way of measuring this force. We need only place a scale between us and the floor, and it will tell us the force that is acting down on us. That is what weight is, and that is also why the correct units for weight are not the same as those for mass. To be correct,

Newton's Apple

Let us consider the force of gravity acting on a falling apple that has a mass of, let us say, about 102 grams.

acceleration due to gravity

Force (in newtons) = mass (in kg) x acceleration due to gravity (in m/s^2), so F = 0.102 x 9.80665 = 1 newton.

weight should be expressed in pounds-force or newtons, because it is a force. It is dependent upon the acceleration due to gravity, which mass is not. The reason that we can use scales as a gauge of mass is that we calibrate them so that the downward force produced by, for example, a mass of one kilogram gives a reading of one kilogram. Under standard conditions on the earth's surface, a mass of one pound weighs one pound-force, and a mass of one kilogram weighs 9.8 newtons.

If the scale were taken to a planet that has a different mass, or a much larger or smaller planet whose surface was further away from its core or much closer to it, the acceleration due to gravity would be different and our one-pound or one-kilogram mass would produce a totally different reading on the scale.

Weightlessness

Weight, then, is a product of mass and acceleration due to gravity. The weight that we experience, or 'perceived weight', results from the force exerted on us by a support that prevents us from accelerating. So what happens when that support is removed? The answer is that, while our weight remains the same, we experience weightlessness. While in free fall under the influence of gravity (in the absence of air resistance), a body experiences no forces acting upon it, and this situation can be created in an airplane that is accelerating downward with the acceleration of gravity. By the same token, if a body is propelled upward with the same acceleration as gravity, it will experience twice its normal weight.

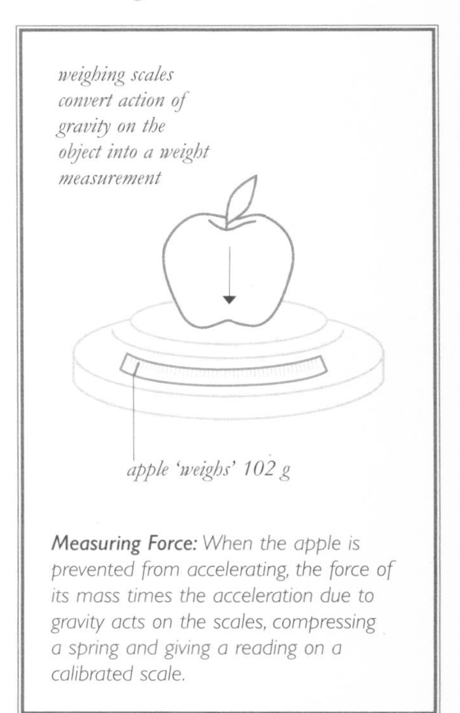

Measuring Force: *When the apple is prevented from accelerating, the force of its mass times the acceleration due to gravity acts on the scales, compressing a spring and giving a reading on a calibrated scale.*

Torque and Leverage

Archimedes is reputed to have said that if he had a fulcrum and a long enough lever, he could move the world. That requires quite some force, but, putting aside a few practical problems, he was theoretically right.

Rotational Force

So far, we have been considering forces acting directly on an object in a straight line, but it is also possible to apply a rotational force to an object, as in the case of using a wrench to turn a nut. In physics, this rotational force is called 'torque' – the product of the length of the wrench (known as the 'moment arm') and the force being applied at right angles to the moment arm. The SI unit of torque is the newton metre (N·m), and the newton centimetre is also commonly used. In the U.S. Customary system, the units of torque are foot pounds-force and inch pounds-force. Although they are usually referred to as 'foot pounds' and 'inch pounds', the units consist of length × force, not mass.

If a force of two newtons is applied at the end of a 3-metre moment arm, the torque is (2 × 3) N·m. If the length of the moment arm is doubled, the torque, likewise, is doubled.

Leverage

The same principle applies in the case of a lever, which can be used to create a 'mechanical advantage'. In a class I lever, the force is applied at one end, the load is at the other end, and the fulcrum about which the lever turns is between the two. The greater the difference between the distance from the applied force to the fulcrum and the distance from the fulcrum to the load, the greater the load that can be lifted. If a fulcrum is placed

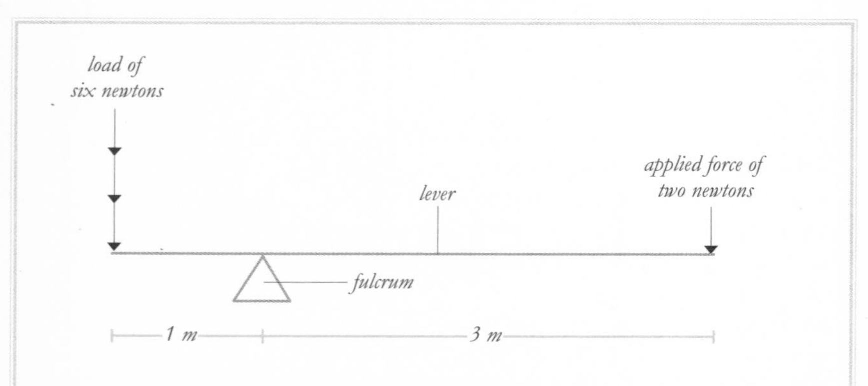

Using Leverage: *When a force is applied at a right angle to a 'moment arm', the force is effectively multiplied by its distance from the fulcrum. A small force applied at the end of a long lever is therefore able to balance a much greater force at a shorter distance beyond the fulcrum, as shown here.*

The Upward Force of Liquids

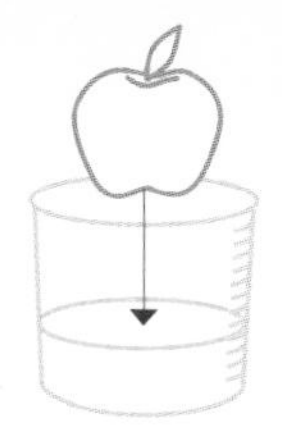

when an object that floats is placed in water . . .

. . . it displaces a volume equal to its own weight

We have seen that the water an object displaces when it is submerged can be used to calculate its volume and, by knowing its mass, we can determine its density. A similar method can be used to measure the weight of an object that floats. Archimedes' Principle states that when a body is totally or partially immersed in a liquid, the upthrust on the object is equal to the weight of the water it displaces. If an object sinks, it weighs more than the weight of the water it displaces; if it floats, the force acting downward on the object (its weight) is balanced by an upward force equal to the weight of the water it displaces. By floating an object, such as an apple, in a full beaker of water and measuring the volume of water that it displaces, we can determine the weight of the apple by multiplying together the volume of water displaced and the density of water.

1 metre from the end of a 4 m lever, and a downward force of two newtons is applied at the end of the longer (3 m) arm, the upward force at the end of the short arm will be six newtons: $f_1 \times d_1 = f_2 \times d_2$, so $f_2 = (f_1 \times d_1)/d_2 = (2 \times 3)/1 = 6$.

Torque and Energy

The newton metre (N·m) is an SI derived unit. Some derived units have their own names in the SI system, and in the next chapter we will see that the joule, a unit of energy, has these same dimensions. It is equivalent to a newton metre. However, by convention, the unit N·m is applied to torque, while joules are applied to energy. The two are not equivalent because in the case of torque the force acts at a right angle to the moment arm, whereas in the case of the joule the force and the distance of movement are in the same direction.

Units of Torque Equivalences

1 foot-pound = 1.3558 newton metres
1 foot-pound = 12.0 inch-pounds
1 inch-pound = 0.1130 newton metres
1 inch-pound = 0.0833 foot-pounds
1 newton metre = 0.7376 foot-pounds
1 newton metre = 8.8508 inch-pounds

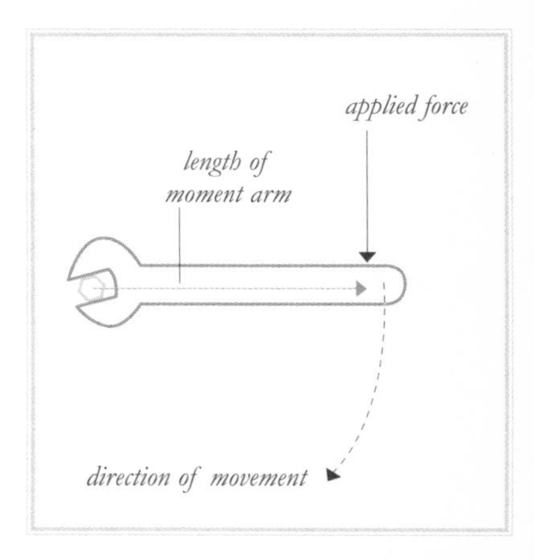

Pressure

FORCE HAS MAGNITUDE AND IT HAS DIRECTION. ANOTHER VARIABLE IS THE AREA OVER WHICH IT IS APPLIED. THE LARGER THE AREA OVER WHICH A SPECIFIED FORCE IS APPLIED, THE SMALLER THE FORCE PER UNIT AREA (OR PRESSURE); THE SMALLER THE AREA, THE GREATER THE PRESSURE.

Units of Pressure

A range of Imperial and U.S. Customary units of force and area have been combined to create units of pressure, but the one most commonly used is the pound-force per square inch. This is abbreviated to psi and often referred to, incorrectly, as pounds per square inch.

In the centimetre–gram–second system, the unit of pressure is the barye (ba), which is equivalent to one dyne per square centimetre ($dyn{\cdot}cm^{-2}$).

The SI derived unit for pressure is the newton per square metre ($N{\cdot}m^{-2}$), but in 1971 this unit was give the special name of the pascal (Pa). Like all SI units, the full range of prefixes can be added to the unit, such as hectopascals and kilopascals (hPa and kPa).

In countries that do not consistently use SI units, other metric and non-metric units of pressure are used in specific contexts.

Spreading the Load

How does area affect pressure? Let us suppose that a person has a mass of 200 pounds. Under standard conditions on the earth's surface, that person will

Pressure Gauge: *This high-pressure gauge is calibrated to indicate pressures up to 4,000 pounds force per square inch (psi) or 28,000 kilopascals (kPa).*

Units of Pressure Equivalences

1 pascal = 1 newton per square metre (1 N/m^2)	1 psi = 1 lbf/in^2
1 pascal = 10^{-5} bar	1 psi = 6,894.76 Pa
1 pascal = 145.04×10^{-6} psi	1 psi = 68.948×10^{-3} bar
1 pascal = 7.5006×10^{-3} mmHg (Torr)	1 psi = 51.715 mmHg (Torr)
1 bar = 10^6 dyn/cm^2	1 Torr = 1 mmHg
1 bar = 100,000 Pa	1 Torr = 133.322 Pa
1 bar = 14.504 psi	1 Torr = 1.33322×10^{-3} bar
1 bar = 750.06 mmHg (Torr)	1 Torr = 19.337×10^{-3} psi

Measuring Pressure

Pressure in a fluid or gas can be measured using a manometer, a device that indicates the pressure as the height of a column of liquid (usually mercury or water) that the pressure supports. Historically, a mercury-filled 'sphygmomanometer' has been used to measure blood pressure and, despite the introduction of sophisticated digital equipment, blood pressure (usually the pressure in the large artery in the arm) is still expressed in millimetres of mercury (mmHg). This leads to figures such as 120/80, meaning a systolic pressure of 120 mmHg and a diastolic pressure of 80 mmHg. ('Systolic' means the maximum arterial pressure occurring during contraction of the left ventricle of the heart, and 'diastolic' indicates the arterial pressure during the interval between heartbeats.) The unit of 1 mmHg is called the Torr, named after

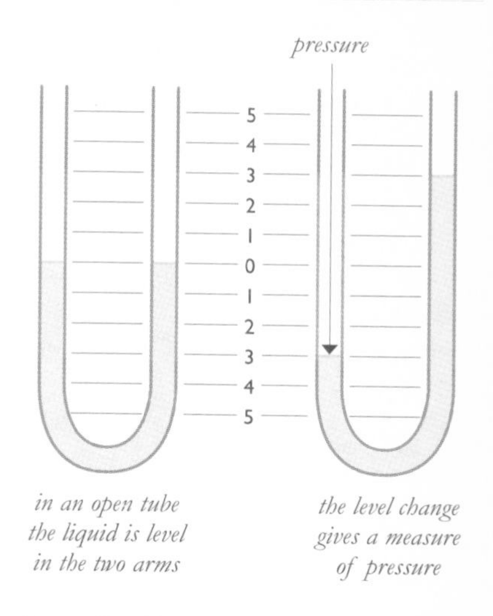

Evangelista Torricelli, the inventor of the barometer (see page 120).

A similar method, using water in a vertical tube, can be used to measure lung pressure by recording the pressure that a person can exert by blowing.

exert a downward force of 200 pounds-force. If the sole area of that person's shoes is a total of 100 square inches, they will exert a pressure of 200/100 = 2 pounds-force per square inch – enough to break through the surface of firm snow. If, on the other hand, the person is wearing snowshoes with an area of 500 square inches, the pressure exerted will be only one-fifth as great (200/500 = 0.4 lbf/in^2) and the snow will support the person.

The same logic explains why people with hardwood flooring tend to be less welcoming to guests in stiletto heels.

Tire Pressure

The required pressure in a vehicle tire is usually expressed in psi (see above) or bar. 1 bar = 1,000,000 dynes per square centimetre, and is almost equal to atmospheric pressure (see page 120). It is not an SI unit, but the bar and the millibar are grudgingly accepted for use with SI units.

When used in the context of vehicle tires, the pressure given is 'gauge pressure' rather than absolute pressure – it refers to the difference between atmospheric pressure and the pressure inside the tire.

Atmospheric Pressure

Air exerts pressure on every surface, and this is referred to as atmospheric pressure. The phenomenon of atmospheric pressure is due mainly to the weight of gas molecules stacked above any given point, and it therefore decreases with altitude. It also varies with temperature, cold air being denser than warm air.

Standard Atmospheric Pressure

In SI units, atmospheric pressure is expressed in pascals, and 'standard atmospheric pressure' (taken to be the typical pressure at sea level at the latitude of Paris) is 101,325 Pa. In chemistry and physics, when certain laws and definitions specify 'at standard temperature and pressure (STP)', this is the pressure being referred to. This 'standard atmosphere' (or atm) is a (non-SI) unit in its own right.

The Mercury Barometer

Expressing pressure as the height of a column of mercury stems from the mercury barometer, which was invented by the Italian scientist Evangelista Torricelli in 1643, and is still use today.

A glass tube with a sealed end is filled with mercury and then up ended in a bowl of mercury. The mercury will sink in the tube until the weight of mercury in the tube is balanced by the pressure of the atmosphere acting on the surface of the bowl of mercury. At standard atmospheric pressure, the column of mercury is about 29.92 inches (760 mm) high at 0 °C. The temperature has to be specified, because the volume of the liquid increases with temperature, and the density therefore decreases, affecting the height of the column.

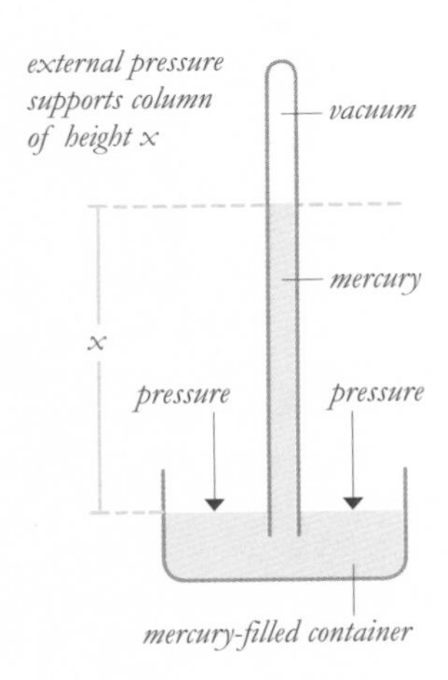

Other liquids can be used in a barometer, but the high density of mercury makes the column conveniently short. Atmospheric pressure will support a column of water approximately 33 feet (10 metres) high.

Meteorological Units

Atmospheric pressure is an important indicator of weather conditions – fair weather being associated with high pressure, and storms with low pressure. The pressure is measured using a barometer, and is often referred to as barometric pressure (the word *bar* is derived from the Greek for 'weight'). In many of the countries that have now adopted the SI system, meteorologists previously used the millibar (1×10^{-3} bar) as the unit of measurement, and they still do in the U.K. Since this unit is based on the dyne and the square cm, it has a direct relationship with the pascal. 1 bar = 100,000 Pa, so one millibar is equivalent to one hectopascal (hPa), and this is the unit found in European weather forecasts – virtually the only context in which it is used.

In Canada, forecasters use the kilopascal (kPa). Barometric pressure can also be expressed in millimetres (mmHg) and inches of mercury (in-Hg), the unit used by the U.S. National Weather Service to report air pressure at the earth's surface.

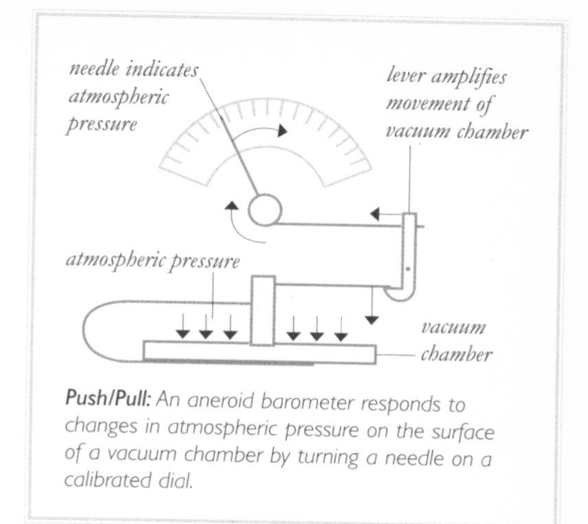

Push/Pull: *An aneroid barometer responds to changes in atmospheric pressure on the surface of a vacuum chamber by turning a needle on a calibrated dial.*

Standard Atmosphere Equivalents

The Standard Atmosphere (atm) is equivalent to the following:

101.325 kilopascals (kPa)
1013.25 hectopascals (hPa)
1.01325 bar
1013.25 millibars (mbar or mb)
14.696 pounds-force per square inch (psi)
29.9213 in-Hg at 0 °C
760 mmHg (Torr) at 0 °C

Height and Depth

Since atmospheric pressure is created by the column of air above the measuring point and decreases with altitude (by about 1 hPa every 8.23 m at low levels), air pressure can be used to calculate the height at which an aircraft is flying. An altimeter is, in fact, a barometer calibrated to display altitude.

Like air, water exerts a pressure due to the weight of water above the measuring point. A diver 33 feet (10 metres) below the surface experiences twice atmospheric pressure, and the pressure continues to increase by approximately one atmosphere for every 33 ft (10 m) of depth. As with the altimeter, the pressure indicates distance from the surface, and a scuba diver's depth gauge is in fact a pressure gauge. The relationship between depth and pressure is linear because water is almost incompressible and therefore its density does not change with pressure (unlike that of air).

CHAPTER TEN

a measure of energy and power

Everything that moves or is under the influence of a force possesses energy. Energy not only makes things happen, it may ultimately be what everything is made of – the very essence of existence. Little wonder, then, that units of energy are used in every branch of science.

'Not only will atomic power be released, but someday we will harness the rise and fall of the tides and imprison the rays of the sun.'

Thomas Alva Edison, Inventor

Force in Action

A FORCE, AS WE HAVE SEEN, IS THAT WHICH PRODUCES A CHANGE OF MOTION IN A BODY. WHEN A FORCE IS APPLIED TO A BODY CAUSING IT TO MOVE IN THE SAME DIRECTION AS THE FORCE, WORK IS DONE.

Doing Work

The amount of work done is the product of the force acting in the direction of motion and the distance moved. To take an example, when a force of 50 pounds-force (50 lbf) moves through a distance of 5 feet in the direction of the force, the actual amount of work done is described as 250 foot-pounds-force (250 ft·lbf).

Using Energy

Doing work uses energy. Energy is the capacity for doing work, and the same units can be used for both. In order to do 250 ft·lbf of work, 250 ft·lbf of energy must be expended. Note that we have defined energy by what it does or can do, rather than saying what it is. This is because we do not know what energy is, except that it is a quality of matter that is capable of bringing about change.

Examples of Kinetic Energy

Motion energy: We started by discussing a mass being moved a distance. This is an example of motion energy – the movement of matter (solid, liquid, or gas) from one place to another as the result of a force being applied.

Heat energy: Also called thermal energy, this is the internal energy (the movement of atoms and molecules) within matter. Any matter that is above absolute zero (see page 84) contains heat energy.

Sound energy: Sound is produced when a force causes matter to vibrate, and sound energy travels in waves of compression by causing the molecules in the material to move back and forth.

Electrical energy: In a sense, this is motion energy on the atomic scale. Electricity is the movement of electrical charges, carried by subatomic particles called electrons, through a conductor as the result of an applied force.

Electromagnetic energy: Also called radiant energy, electromagnetic energy travels in wave form and ranges in wavelength from gamma-rays and x-rays, through visible light and infrared radiation, to microwaves and radio waves. Energy from the sun (solar energy) reaches the earth in the form of radiant energy.

Examples of Potential Energy

Gravitational energy: A mass in a gravitational field, such as a body of water at the top of a mountain, possesses potential energy by dint of its position. This potential energy is converted into kinetic energy when the mass accelerates downward.

raised ball possesses potential (gravitational) energy

rolling ball possesses kinetic energy

Chemical energy: The atoms and molecules in a material are held together by chemical bonds, and energy is stored in these bonds. Some of this stored energy is released in chemical reactions such as the combustion of gasoline. A battery stores chemical energy that can be transformed into electrical energy.

Mechanical energy: Objects under tension or compression, such as springs, store mechanical energy.

Nuclear energy: The particles that compose the nucleus of an atom are held together by energy, some of which is released when nuclear particles are separated or combined (during fission and fusion). Solar energy comes from the fusion of hydrogen nuclei.

Much from Little: *In nuclear fission, a tiny amount of mass is converted into a great deal of energy.*

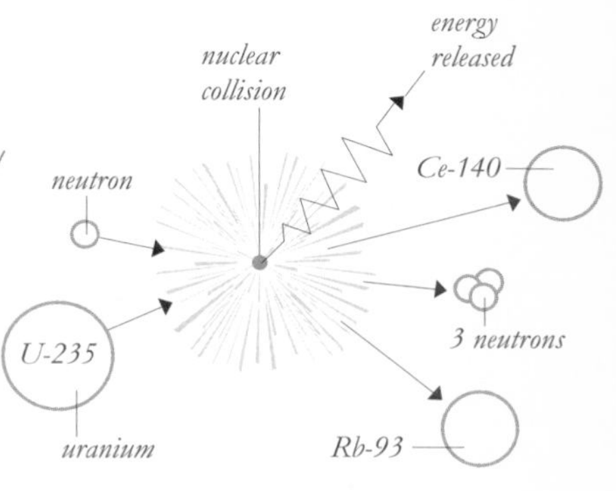

In the case of energy in the form of mechanical work, the change consists of motion, or a change in position, but energy can also bring about changes in size, shape, temperature, density, volume, chemical structure, phase (from solid to liquid, for example), and so on.

In bringing about these changes, energy itself is transformed from one form to another. For example, in a steam engine, heat energy changes into pressure, which pushes a piston, which turns a flywheel that can turn a generator that changes motion into electricity.

Energy can also become mass, and mass can become energy, as Einstein's equation, $E = mc^2$ (Energy = mass x the speed of light squared), tells us.

NON-SI UNITS OF ENERGY AND POWER

SINCE THE IMPERIAL AND U.S. CUSTOMARY SYSTEMS HISTORICALLY SHARE UNITS OF FORCE AND LENGTH, THEY ALSO SHARE SUCH ENERGY UNITS AS THE BTU AND THE FOOT-POUND.

In the pre-SI centimetre–gram–second system, the unit of energy was the 'erg', equal to the work done by a force of one dyne acting over a distance of one centimetre. As we have already seen, one joule is 1 newton x 1 metre. One dyne (the force required to accelerate a mass of one gram at a rate of one centimetre per second squared) is equal to exactly 10^{-5} newtons and one centimetre is 10^{-2} metres. One erg is therefore 10^{-5} newtons x 10^{-5} metres, making it exactly 10^{-7} joules.

THE WORD 'ERG' IS DERIVED FROM THE GREEK *ERGON*, MEANING WORK. THE WORD ENERGY HAS THE SAME ROOT.

THE FOOT-POUND FORCE

We have seen that the pound force is a unit of force, and since energy is force multiplied by distance it is logical that the Imperial and U.S. Customary unit of mechanical work/energy should be the foot-pound force (ft·lbf), which is approximately equal to 1.356 J. (In the U.S., the joule is only used in scientific contexts.)

We have also seen that power is energy per unit time, and derived units of power are therefore the foot-pound force per minute (ft·lbf/min) and foot-pound force per second (ft·lbf/s). The horsepower is also derived from the foot-pound force.

BRITISH THERMAL UNITS

Sometimes called a heat unit, the British thermal unit (Btu or BTU) is the unit of heat energy used in North America. (In Britain, it has largely been replaced by the joule.) The Btu is defined as the amount of heat required to raise the temperature of one pound of water by one degree Fahrenheit. This, of course, is the amount of heat that James Joule was using as a unit when demonstrating the mechanical equivalent of heat, and the Btu is now equal to approximately 778.169 foot-pounds, only slightly different from the value found experimentally by Joule. One Btu is also equal to approximately 252 calories (see below), 1.055 kilojoules, or 0.293 watt-hours.

The Btu is also sometimes used as a unit of power when describing the capacity of electrical equipment such as heating furnaces. What is actually meant in this case is Btu per hour.

THE CALORIE

Defined in a similar way to the Btu, the calorie (or small calorie) is the quantity of heat required to raise the temperature of one gram of water by 1 °C at standard atmospheric pressure. In fact,

James Watt and the Horsepower

The design improvements that Scottish instrument maker and engineer James Watt made to the steam engine had a major impact on the pace of the Industrial Revolution in eighteenth-century Britain. When he began to market his new steam engines – largely to the mining industry, to pump water from the shafts – he emphasised their efficiency over the older Newcomen engines that his were to replace, and he offered to be paid just on the basis of the coal that his engines saved. When it came to selling his engines to businesses that were using horses, however, he had to find another way in which to be paid, so he compared the power of his engines to their equivalent in horses. He knew that his engines could lift *x* pounds of water *x* feet in a given space of time, so he worked out the power of the horse in foot-pounds force per minute, coming up with a figure of 33,000 ft·lbf/min, or 550 ft·lbf/s. This has been the definition of the horsepower (hp) ever since. One horsepower is roughly equivalent to 746 watts. The horsepower-hour is sometimes used as a unit of energy, and is approximately equivalent to 2.685 MJ.

Donkey Work: *A direct comparison between the power of his engine and that of a horse demonstrates that different forms of energy can be used to do the same work.*

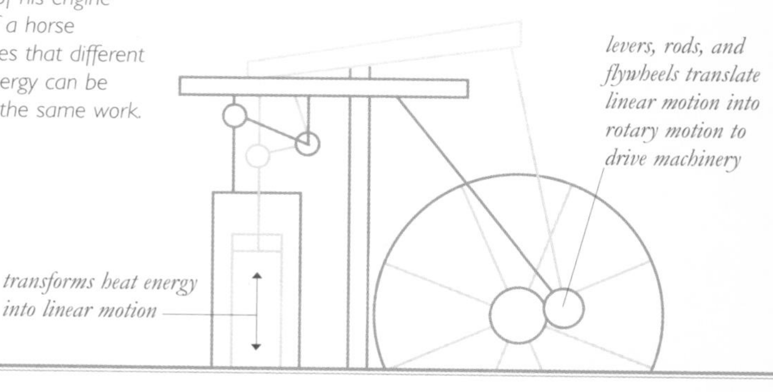

the amount of heat required depends upon the starting temperature of the water, and in consequence there are several definitions and equivalences of the calorie. The 'thermochemical calorie' is defined as 4.184 J exactly.

The large calorie, Calorie, or kilocalorie is a unit used in nutrition, and it is the amount of heat required to raise the temperature of one kilogram of water by 1 °C at standard atmospheric pressure, making it equivalent to 1,000 small calories. The kilocalorie (kcal), as well as the joule in European countries, is the unit used to indicate the digestible energy content of packaged food.

SI Units of Work, Energy, and Power

Energy comes in many forms. Historically, many of these have been investigated separately, and this is reflected in the range of units in which the different forms can be expressed.

The SI Unit of Work/Energy

No matter what form it is in, all energy is basically the product of force and distance and is convertible from one form to another. The SI system therefore uses one unit for all kinds of energy. This is the joule (J), the amount of energy needed to push a distance of one metre with a force of one newton. 1 joule = 1 newton-metre (Nm), the same dimensions as the unit for torque. In fact, a torque of one newton-metre moved through one radian does exactly one joule (or newton-metre) of work.

The Mechanical Equivalent of Heat

The joule is named after James Prescott Joule (1818–1889), an English scientist who established, through meticulous measurement, that motion and heat were interchangeable. By measuring the rise in the temperature of a given volume of water when a paddlewheel was turned in it, driven by a falling weight, he found that a given amount of work done by the falling weight always produced the same rise in temperature. As an average figure, he found that the work done by a weight of 772 pounds falling one foot raised the temperature of one pound of water by 1 °F. He had demonstrated not only that heat is produced by motion, but also that when energy is converted from one form to another the total amount of energy in the system is unchanged. His findings, and the independent work of the German physicist Julius Robert von Mayer, led to the theory of the conservation of energy and the First Law of Thermodynamics.

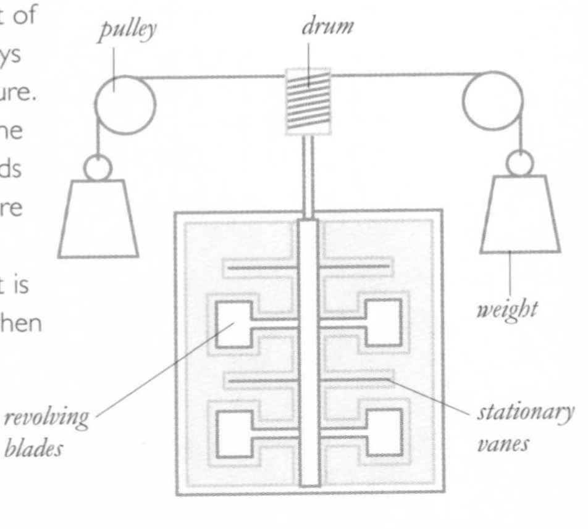

Motion to Heat: *Joule's paddlewheel apparatus demonstrated that movement, or kinetic energy, and heat are interchangeable.*

The Conservation of Energy

The First Law of Thermodynamics states that energy can never be created or destroyed. When energy is transformed or work is done, the total amount of energy in the system remains unchanged. Most transformations result in some dissipation of energy in the form of heat and, overall, energy changes from high grade (energy in the form of motion) to low grade (energy in the form of heat). This is the gist of the Second Law of Thermodynamics. More and more energy becomes unavailable to do work, and this is described as an increase in entropy. It is in this irreversible process that the directionality of the 'arrow of time' can be perceived.

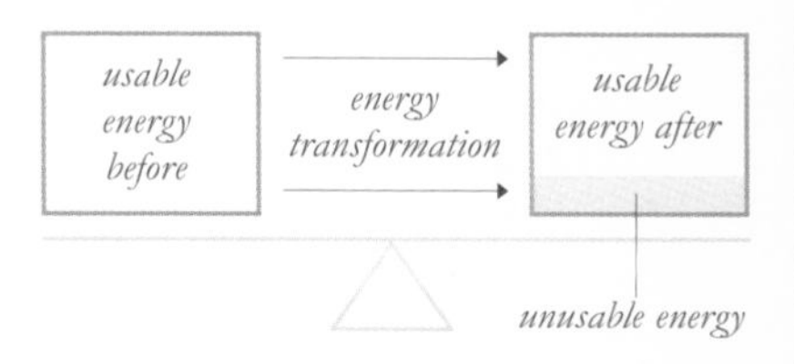

Nothing Lost, Nothing Gained: *When energy is changed from one form to another the total amount of energy remains unchanged.*

THE SI UNIT OF POWER

Power is a measure of the rate at which work is being done or energy is being used, expressed as the amount of work/ energy per unit time. In the SI system, the derived unit of power is the joule per second, and it is given the special name of the watt (W), after the Scottish engineer James Watt. If 100 joules of work are done in one second (using 100 joules of energy), the power is 100 watts.

Electrical power is commonly expressed in watts and kilowatts (kW) at the household level, and in megawatts (MW) when discussing power generation and distribution. A typical incandescent light bulb uses energy at the rate of 40 to 100 watts, while an electric water heater may be rated at 6 kilowatts. A nuclear power station may generate several hundred megawatts of electricity.

The watt is a convenient unit in this context, because the power in an electrical circuit, as expressed in terms of watts, is equal to the voltage in volts multiplied by the current in amperes (see page 132).

THE KILOWATT-HOUR

The quantity of any kind of energy can be expressed in joules, but the kilowatt-hour is commonly used to express quantities of electrical energy (for example on utility bills). One kilowatt-hour is one kilowatt of power for 60 minutes. This is not an SI unit, despite the fact that the kilowatt is, because the hour is not an SI unit.

Power is energy per unit time, so power multiplied by time is a measure of energy. Watts are joules per second, so a watt-second is a joule. Since a kilowatt-hour is one thousand watt-seconds x 3,600, it is equivalent to 3.6×10^6 joules, or 3.6 megajoules (3.6 MJ).

Radioactivity

Atoms are composed of a positively charged nucleus and negatively charged electrons orbiting the nucleus. The nucleus itself consists of positively charged protons, which repel each other with a strong electromagnetic force. The strong nuclear force overcomes this and holds them together, but as a result of this conflict, many isotopes are unstable. When a nucleus fragments, it emits radiation, including alpha and beta particles, and gamma radiation. Due to the strength of the forces at work and the tiny size of the nucleus, the energy released by this process – known as radioactive decay – is considerable.

Measuring Radioactivity

Each radioactive isotope decays at a different rate, and this is expressed as the 'half-life' of the isotope. The half-life is basically a measure of the time taken for one half of the atoms in a sample to decay, and this rate is a constant for any particular isotope.

Radioactivity is measured in terms of the number of individual decay events taking place in a given sample every second. It is, of course, dependent upon the quantity of radioactive material present and the half-life of the isotope.

The original unit of radioactivity was the curie (named after Marie Curie, a pioneer in the science of radioactive material), and it was defined as the radioactivity of one gram of pure radium.

In the SI system, the unit is individual decay events per second, giving a unit of s^{-1}. For the sake of clarity, the unit in this context is given the special name of the becquerel. This is a very small unit, and the gigabecquerel is more commonly used. One curie is the equivalent of 37 gigabecquerels (GBq).

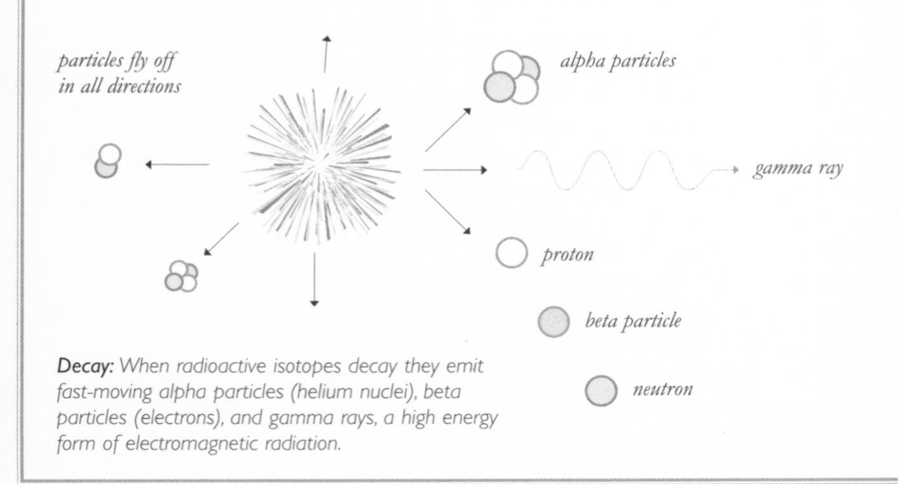

Decay: *When radioactive isotopes decay they emit fast-moving alpha particles (helium nuclei), beta particles (electrons), and gamma rays, a high energy form of electromagnetic radiation.*

Henri Becquerel

The SI unit of radioactivity is named after Henri Becquerel, a French scientist who, in 1896, discovered the spontaneous emission of nuclear radiation when he noticed that uranium salts caused a photographic plate to become fogged without the plate being exposed to light. In 1903, Becquerel, Marie Curie, and her husband Pierre shared the Nobel Prize in Physics for their work on radioactivity (see page 93).

Ionizing Radiation

All nuclear radiation is described as 'ionizing'. This means that all the particles and electromagnetic radiation emitted by a decaying nucleus possess enough energy to remove an electron from its orbit, causing an atom or molecule to become electrically charged, or ionized. This is important because ionization can destroy living tissue and can damage DNA, potentially causing mutation or inducing cancer.

For this reason, it is vitally important to be able not only to measure the radioactivity of a sample of material, but also to be able to monitor the amount of radiation that a person receives and to evaluate its potentially harmful impact.

On the positive side, the destructive power of ionizing radiation can be used to target and destroy cancerous cells, and this is the purpose of radiotherapy. Again, it is essential to be able to prescribe and administer the correct dose of radiation.

Units of Absorbed Dose

The damaging aspect of ionizing radiation is the energy that it deposits in the irradiated material (such as human tissue). The greater the amount of energy, and the smaller the amount of material that absorbs it, the higher the dose, so absorbed dose is expressed as energy per unit mass.

The pre-SI unit of radiation dose is the rad (standing for 'radiation absorbed dose'). It is a centimetre-gram-second unit, defined as 100 ergs per gram. This unit is no longer used anywhere but the United States, where the National Institute of Standards and Technology discourages its continued use.

In the SI system, in which the unit of energy is the joule and the unit of mass is the kilogram, the unit of absorbed dose is the joule per kilogram, but to avoid confusion, especially with the unit of equivalent or effective dose (see below), this unit is given the special name of the gray (Gy), named after the English physicist and radiologist Louis Harold Gray. 1 gray = 1 joule of energy absorbed per kilogram of tissue = 100 rad.

Knowing the absorbed dose is not enough to tell us what the biological effects will be, because different types of radiation affect the tissue differently, and different types of tissue differ in their sensitivity to radiation. To give a better indication of the likely biological effect of a given dose of radiation, a measurement called the dose equivalent, which takes these factors into account, is used. It, too, has units of joules per kilogram, but the special name of sievert (Sv) is given to this unit to avoid confusion.

Units of Electricity

When it comes to electricity, there is, surprisingly, no difference between the Imperial, U.S., and SI units. This is because the fundamentals of the metric system were already in place when electricity was being investigated.

The electrical measurement system therefore constitutes a coherent whole, and the base units on which it is founded are at the heart of SI.

The Ampere

The unit of electric current is one of the seven SI base units: the ampere (A). The Ampere is 'that constant current which, if maintained in two straight parallel conductors (wires) of infinite length, of negligible circular cross-section, and placed 1 metre apart in vacuum, would produce between these conductors a force equal to 2×10^{-7} newton per metre of length.'

The Coulomb

Each electron has a charge, but as a unit of measurement this would be

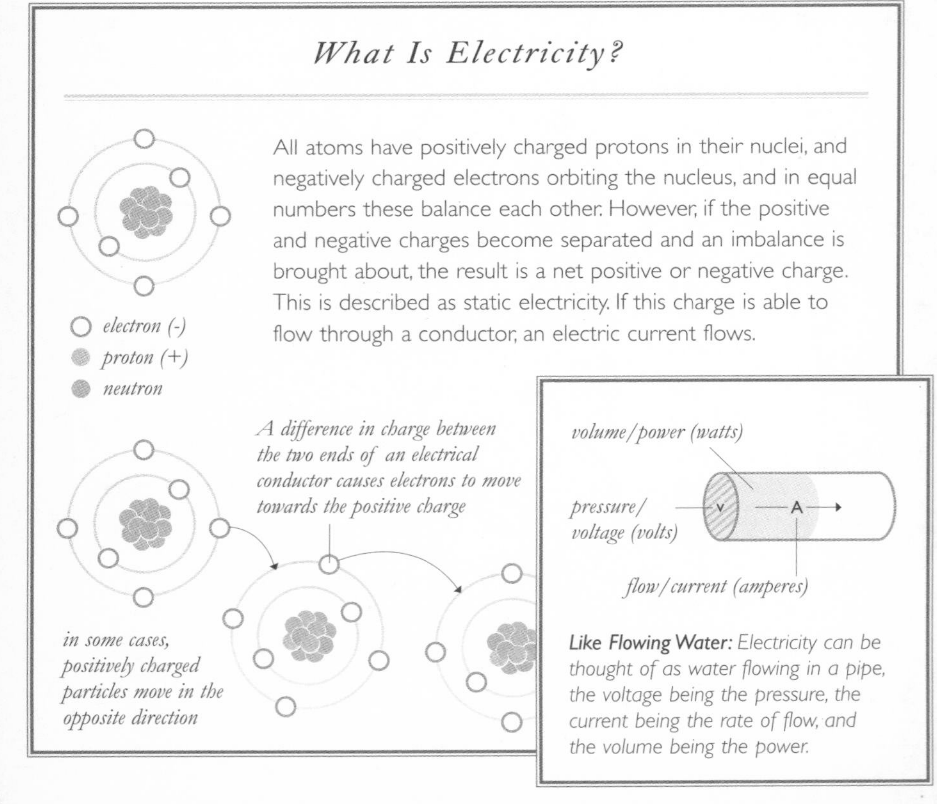

ridiculously small, so instead the unit of charge is the coulomb. It is defined as the amount of charge that passes along a wire when one ampere flows for one second. A coulomb of negative charge is that of 6,280,000,000,000,000,000 (6.28×10^{18} electrons).

The Volt

Voltage is effectively the force pushing the charge through the conductor, also called the potential difference or electromotive force (EMF). The volt is the force that will give 1 coulomb of electrons 1 joule of energy.

The Ohm

Except at temperatures approaching absolute zero (see page 84), all materials have some resistance to the flow of an electric current. The ohm (Ω), the unit of resistance, is that resistance which requires a volt of EMF to drive one ampere of current.

The Siemens

Resistance (R) is a measure of the difficulty with which electricity will flow in a conductor. Conductance is the reciprocal of this – the ease with which electricity will flow. The unit of conductance is the siemens, defined as 1/Ω, and previously called the mho.

The Watt

As we have seen, the watt (W) is the SI unit of power, equivalent to a joule per second. One volt of EMF across a one-ohm resistance will cause one ampere of current to flow and will produce one joule of heat energy per second (the conversion of electrical energy into heat energy is 100 percent efficient, and there is no loss). To put it another way, the watt is the amount of power required to push a coulomb of charge (one ampere) against an EMF of one volt for one second.

The Farad

A battery stores electrical charge in a chemical form, but charge can also be stored in a capacitor, which consists of two plates with an insulative layer between them. Capacitance is a measure of the ratio between the charge held, in coulombs, and the potential difference between the plates, in volts. The unit of capacitance is the farad (F). For most electronic circuitry this is too large a unit, and the microfarad (μF), nanofarad (nF), and picofarad (pF) are more commonly used.

The Weber

A change in magnetic field will produce an electric potential in a conductor. This change, or magnetic flux, is measured in webers (Wb), and a change of magnetic flux of one weber per second will induce a potential of 1 volt.

The Tesla

The density of a magnetic field, also known as the magnetic flux density, is measured in tesla (T), which is one weber per square metre.

The Henry

Inductance is a measure of the electrical potential that is induced in a circuit by a change in current, and the unit of inductance is the henry (H). If a circuit has an inductance of one henry, a change of one ampere per second produces an EMF of one volt.

Radiant Energy

The electromagnetic spectrum extends from radio waves (low frequency) to gamma rays (high frequency), and all forms of electromagnetic radiation carry energy that is passed on to material with which it interacts.

Electromagnetic radiation can be considered as having a wave form, with electric and magnetic waves oscillating at right angles to each other, or as a stream of particles called photons. The duality of this wave/particle character is resolved in quantum field theory.

Energy and Power

The energy carried by any form of electromagnetic radiation can be expressed in the SI unit of joules, and the energy per unit time, or power, has the predictable unit of watts. The intensity of the power can be expressed in various ways using units such as watts per steradian (see opposite), or watts per square metre.

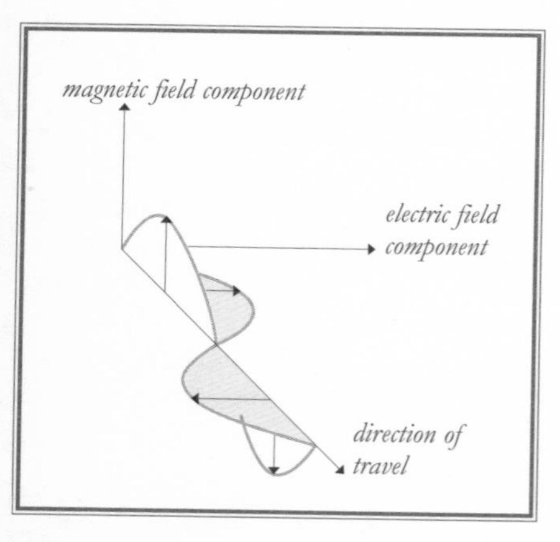

Wave Form: *Electromagnetic radiation takes the form of two sets of waves – one electrical and one magnetic – at right angles to each other, with the same frequency and wavelength.*

Velocity, Wavelength, and Frequency

The speed of all electromagnetic radiation in a vacuum is, as light (see page 102), a constant 299,792,458 metres per second.

The wavelength of any given type of radiation is the distance between repeating units, and is usually taken as the distance between successive peaks or successive troughs. The Greek letter *lambda* is used to denote wavelength. The units of wavelength are metres, ranging from picometres in the case of gamma radiation, through to megametres for some of the longest radio waves. Visible light lies between about 400 and 700 nanometres.

The frequency of a wave form is the number of waves per unit time, so the SI unit is one per second (1/s). This is the same as the becquerel, and so to avoid confusion the unit was given the special name of hertz (Hz), named after the German physicist Heinrich Rudolf Hertz. The relationship between the speed (v), wavelength (l), and frequency (f) is f = v/l. Since the speed of the radiation in any given medium is constant, frequency and wavelength are inversely proportional. The shorter the wavelength, the higher the frequency.

The Steradian

We have already encountered the two-dimensional radian, an SI derived unit, in the context of angular velocity. The 'steradian' is the radian's three-dimensional brother, and it is the solid angle subtended at the centre of a sphere whose radius is r by an area of r^2 on the surface of the sphere. Since the surface area of a sphere is $4\pi r^2$, the surface measures 4π steradians. The steradian, too, is an SI derived unit, with the dimensions of $m^2 \times m^{-2}$, or 1, but it has been given this special name.

r^2

r

Squared Radian: *If the area that the steradian subtends is thought of as a square, the length of each side of that square is one radian.*

Light: A Special Case

Given that light is just another form of electromagnetic radiation, no special units are needed in order to measure it, but light is a very special case simply because of the importance of our sense of sight. For this reason, and the everyday practical use that we make of this form of radiation, it has its own dedicated units in the SI system, including its own base unit.

The Candela

At the start of the twentieth century, light intensity was measured in candlepower, based on a flame or incandescent filament, but this was improved in 1948 when the CGPM ratified a new unit called the candela, based on the radiation from freezing platinum. If that sounds cold, bear in mind that platinum melts at 1768.3 °C (2041.4 K).

The difficulties of recreating the standard, combined with advances in the accuracy of radiation measurement, led the 16th CGPM to adopt the current definition, which is derived from the watt, in 1979:

'The candela is the luminous intensity, in a given direction, of a source that emits monochromatic radiation of frequency 540×10^{12} hertz and that has a radiant intensity in that direction of $1/683$ watt per steradian.'

Electromagnetic radiation is normally specified by wavelength, not frequency as it is here, but the wavelength changes depending on the medium through which it is passing, which frequency does not.

The Lumen and the Lux

The candela is a unit of measurement of the brightness of a source in any one direction, but this does not tell us the total amount of light being emitted, the luminous flux. For this, the unit of the lumen is used – equivalent to candela per steradian.

To measure the light energy falling on an object, the unit of lumens per square metre is used. This is called the lux.

Measuring With Lasers

Light, then, consists of electromagnetic radiation. The various colours that we perceive are the result of the radiation being at various wavelengths. White light contains all the visible wavelengths, and this can be demonstrated by splitting a beam of white light by passing it through a prism, which separates the radiation of different wavelengths and reveals the colours of the rainbow.

The light produced by a laser (which stands for Light Amplification by Stimulated Emission of Radiation) is special in several respects. The light is of one single wavelength, it is coherent (meaning that the waves are all in phase and are parallel), and it can be extremely powerful.

white light hits prism

the various wavelengths of which white light is composed are refracted by different amounts, spreading them out

Newton's Colours: *Sir Isaac Newton classified the colours of the rainbow as red, orange, yellow, green, blue, indigo, and violet, but they actually form a continuum.*

How a Laser Works

The electrons of any particular atom or molecule exist at specific energy levels, which can be imagined as orbits around the nucleus. When light hits an electron it can raise the electron's energy level by a specific amount and cause it to jump to a higher-level orbit. When the electron drops back to its original level, a packet of light energy, or photon, is released, and the wavelength (or colour) of the light is specific to the material involved. Ruby (which is an aluminium oxide containing chromium), for example, emits red light.

A laser consists of a rod of the 'lasing' material surrounded by a high-intensity lamp. At one end of the rod there is a mirror, and at the other is a partial mirror (one that reflects some light and lets some light through).

In the case of a ruby laser, a flash of light from the lamp excites the electrons in the chromium atoms to a higher energy level, and they release this energy in all directions as photons of a specific wavelength – that of red light. Light that is travelling along the length of the rod is reflected back into the rod by the mirrors, striking other chromium atoms and causing them, too, to emit light. High-energy, single-colour light is built up in this way until it is emitted as a coherent pulse from the laser.

Units of Laser Power

Lasers have many uses, from reading CDs to starting nuclear fusion reactions, and the energy they can generate is an important quality. In the early days of laser development, pioneer laser physicist Theodore Maiman (b. 1927) came up with a way of measuring relative laser power by seeing how many Gillette razor blades the laser could burn a hole through. Since the thickness of these blades was fairly uniform, this provided a useful scale, enabling researchers to evaluate performance in 'Gillettes'. Early prototypes were in the 2–4 Gillette range.

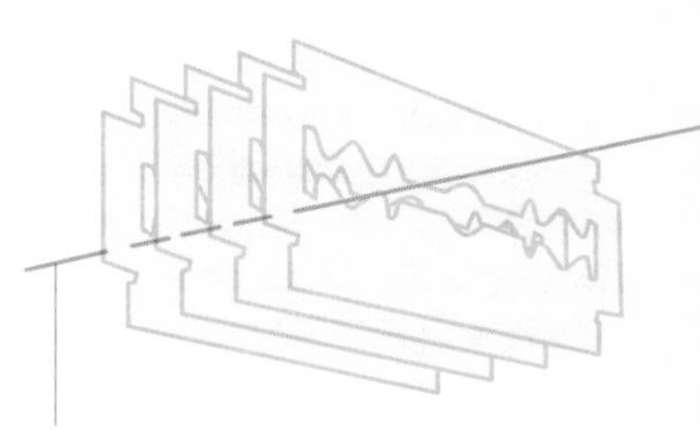

the number of blades through which a laser would cut provided an early non-standardised measure of its power

Interferometry

Interference occurs when waves interact, be they waves on the surface of water or waves of electromagnetic radiation. If two waves reach the same point at the same time and they become synchronized (i.e. their peaks reach the same point at the same time), constructive interference takes place and the amplitude, or size, of the waves is increased. They effectively reinforce each other.

On the other hand, if they are 'out of synch' and the peak of one coincides with the trough of the other, destructive interference occurs and they cancel each other out.

Lasers can be used to exploit this phenomenon because the light they emit is coherent. If a laser beam is split and the two halves of the beam are recombined after being reflected along different paths, the interference between them will show whether the two paths differ by even a fraction of a wavelength.

Measuring Distance

Ultrafine methods of measuring time mean that the length of time taken for a laser pulse to reach a target and be reflected back to the emitter can be used to give an accurate measurement of the distance travelled, and hence the distance to the target.

Measuring Speed

A lidar (light detection and ranging) speed gun uses a laser to detect the distance to a moving car by taking measurements several hundred times in a second. The rate of change of this distance tells police whether the car is exceeding the speed limit.

Explosive Energy

When the SI units kiloton or megaton are used to describe an explosion, the units are not being used to express weight, but an energy equivalent.

The destructive effect of an explosion is proportional to the energy released in the form of heat and the work done by the expanding gases generated by the explosion. As such, the strength of an explosion, or the potential strength of an explosive device, can be expressed in joules or other units of work/energy. However, for historical reasons, and because the joule is a rather small unit to use for a modern thermonuclear device, the energy of an explosion or explosive device is expressed in TNT equivalent units in many contexts, including nuclear weapon-control treaties.

TNT, or trinitrotoluene, is a conventional explosive that, for arms-control purposes, is taken to have 1,000 thermochemical calories (4.184 kilojoules) per gram. The units of kiloton (kt), megaton Mt), and even gigaton are commonly used not only to describe conventional and nuclear warheads but also the energy released by the impact of asteroids. The explosive strength of a one-megaton nuclear weapon, for example, is equivalent to that of one million tons of TNT. The nuclear bombs dropped on Hiroshima and Nagasaki had yields of about 15 and 25 kilotons respectively. The USSR's Tsar Bomba, the largest nuclear weapon ever tested, had a yield of 50 megatons, 2,000 times greater than the bomb dropped on Nagasaki.

1 ton (1 megagram)	TNT equivalent	$= 4.18 \times 10^{9}$ joules (4.18 gigajoules)
1 kiloton (1,000 tons)	TNT equivalent	$= 4.18 \times 10^{12}$ joules (4.18 terajoules)
1 megaton (1,000,000 ton)	TNT equivalent	$= 4.18 \times 10^{15}$ joules (4.18 petajoules)

Asteroid Impact?

In June 1908, an explosion took place in the almost uninhabited Tunguska region of central Siberia that flattened more than 2,000 square miles of forest. Preceded by a bright streak of light, seen more than 300 miles away, and followed by a shock wave that was felt over 600 miles away, the explosion is estimated to have had a strength of between 10 and 20 megatons, and to have been caused by the impact of a chunk of comet.

Wind Power

The Beaufort Wind Scale was created in 1805 by the British naval officer Sir Francis Beaufort to help sailors estimate the speed of the wind on the basis of its observed effects, originally on the sails of a ship. With the introduction of steam ships, the Beaufort scale was linked to the effects of the wind on the sea itself. The Beaufort scale is still used today to express wind strengths, most notably in the British shipping forecasts broadcast on BBC Radio 4.

Force	*Knots*	*Classification*	*Appearance of Wind Effects*
0	less than 1	Calm	SEA: Surface smooth and mirror-like LAND: Calm, smoke rises vertically
1	1–3	Light Air	SEA: Scaly ripples, no foam crests. LAND: Smoke drift indicates wind direction, still wind vanes
2	4–6	Light Breeze	SEA: Small wavelets, crests glassy, no breaking LAND: Wind felt on face, leaves rustle, vanes begin to move
3	7–10	Gentle Breeze	SEA: Large wavelets, crests begin to break, scattered whitecaps LAND: Leaves and small twigs constantly moving, light flags extended
4	11–16	Moderate Breeze	SEA: Small waves 1–4 ft becoming longer, numerous whitecaps LAND: Dust, leaves, and loose paper lifted, small tree branches move
5	17–21	Fresh Breeze	SEA: Moderate waves 4–8 ft taking some spray longer form, many whitecaps LAND: Small trees in leaf begin to sway
6	22–27	Strong Breeze	SEA: Larger waves 8–13 ft, whitecaps common, more spray LAND: Larger tree branches moving, whistling in wires
7	28–33	Near Gale	SEA: Heaps up, waves 13–20 ft, white foam streaks off breakers LAND: Whole trees moving, resistance felt walking against wind
8	34–40	Gale	SEA: Moderately high (13–20 ft) waves begin to break into spindrift, of greater length, edges of crests, foam blown in streaks LAND: Whole trees in motion, resistance felt walking against wind
9	41–47	Strong Gale	SEA: High waves (20 ft), sea begins to roll, dense streaks of foam, spray may reduce visibility LAND: Slight structural damage occurs, slate blows off roofs
10	48–55	Storm	SEA: Very high waves (20–30 ft) with overhanging crests, sea white with densely blown foam, heavy rolling, lowered visibility LAND: Seldom experienced on land, trees broken or uprooted, 'considerable structural damage'
11	56–63	Violent Storm	SEA: Exceptionally high (30–45 ft) waves, foam patches cover sea, visibility more reduced
12	64+	Hurricane	SEA: Air filled with foam, waves over 45 ft, sea completely white with driving spray, visibility greatly reduced

CHAPTER ELEVEN

a measurement miscellany

Since time immemorial, practitioners of each trade, art, or science have striven to create their own private world of knowledge. What better way could there be to preserve the secrets of a craft than having its own units and methods of measurement?

'The art of measurement, by showing us the truth, would have brought our soul into the repose of abiding by the truth, and so would have saved our life.'

Protagoras, Greek Philosopher

UNITS OF DIGITAL INFORMATION

THE SPOKEN WORD, WHICH CONSISTS OF CONTINUOUS VARIATIONS, IS ANALOGUE INFORMATION. BY CONTRAST, DIGITAL INFORMATION IS MADE UP DISCRETE ELEMENTS RATHER THAN A VARYING CONTINUUM.

The term 'digital' is most commonly applied to information stored in or manipulated on a computer, but Morse code, semaphore, braille, and even smoke signals are all made up of a limited number of repeated elements – dots, dashes, spaces, flag positions, raised dots, or puffs of smoke – and are therefore digital.

The digital elements on which computers operate are binary. In the binary system, each individual digit is either a 0 or a 1, and the word 'bit' is an abbreviation of 'binary digit'.

In this binary language, 8 bits provide 256 different combinations. A parcel of 8 bits is called a byte (the French for byte is *octet*), and this is the common unit of file size. For example, the number 122 is represented by the 8-bit string 01111010 in binary notation, as shown here:

128	64	32	16	8	4	2	1
0	1	1	1	1	0	1	0

64 + 32 + 16 + 8 + 2 = 122

In addition, this binary number represents the letter Z in ASCII code – the American Standard Code for Information Interchange, which is used to represent the letters and symbols of the computer keyboard.

Being a binary system, the right-hand column represents single units and the value of a unit in each column to the left is 2 x the value of a unit in the column before it. (In our everyday decimal, or base 10, system, a unit in any column has 10 times the value of a unit in the column to its right.)

Through the use of computers and software, scanners and other hardware, most forms of information, including text, numerical data, images, and recorded sound can be created in, or converted into, a digital form and then manipulated, stored, copied, printed, or written to other forms of data storage. They can also be emailed, posted on a website, and so on.

Boolean Algebra

English mathematician George Boole (1815–1864) was a remarkable man. By the age of 19 he was running his own school, and at the age of 34 he was appointed to the chair of mathematics at Queens College, Cork – all despite having little formal education. In an attempt to formulate a rational basis for human thought, he developed a form of logic – Boolean algebra – that is now the foundation on which all digital computers operate.

Kilobytes and Megabytes

After all that we have seen concerning the use of SI prefixes and the way that the system works, you would expect to be fairly safe in assuming that a kilobyte is 1,000 bytes. Wrong! Because the binary system is base 2, the units increase by powers of 2: 128, 256, 512, and 1,024. It's close to 1,000, but it is actually 2^{10}, not 10^3. The difference is only 2.4 percent, but a 'binary' yottabyte (2^{80} bytes) is more than 20 percent greater than its SI definition of 10^{24} bytes. There is a move already taking place to replace this misuse of the SI prefixes with new binary prefixes to avoid confusion, and in some fields these are already in use. The actual values of the megabyte and its larger colleagues are shown below, together with the proposed binary names and their symbols.

Correct Metric Uses

Apart from the fact that the use of the SI prefixes is inaccurate, there is another reason to sort out the situation. Because they are made up of bytes, file sizes need to be expressed in binary numbers. But other aspects of information technology, such as memory size and the rates at which data can be transferred, can be, and are, legitimately expressed in decimal numbers. A computer manufacturer declaring a hard disk to have a capacity of 10 gigabytes is using the term in its 'true' decimal sense, meaning 10 billion bytes (not the binary equivalent of 10.74 billion bytes). Confusion can also arise when telecommunication engineers speak of data transfer rates in megabytes (meaning million bytes) per second, or MBps, while computer network engineers use the same terms to mean binary megabytes.

SI name	*Number of bytes*	*Power*	*Proposed binary name*
Kilobyte	1,024	2^{10}	kibibyte (KiB)
Megabyte	1,048,576	2^{20}	mebibyte (MiB)
Gigabyte	1,073,741,824	2^{30}	gibibyte (GiB)
Terabyte	1,099,511,627,776	2^{40}	tebibyte (TiB)
Petabyte	1,125,899,906,842,624	2^{50}	pebibyte (PiB)
Exabyte	1,152,921,504,606,846,976	2^{60}	exbibyte (EiB)
Zettabyte	1,180,591,620,717,411,303,424	2^{70}	zebibyte (ZiB)
Yottabyte	1,208,925,819,614,629,174,706,176	2^{80}	yobibyte (YiB)

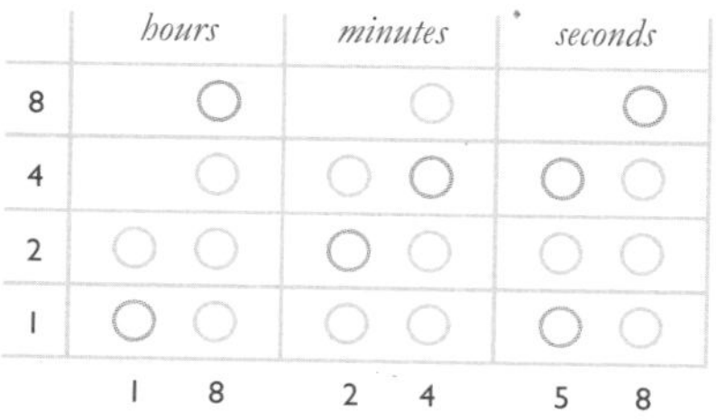

	hours		*minutes*		*seconds*	
8		○		○		○
4		○	○	○	○	○
2	○	○	○	○	○	○
1	○	○	○	○	○	○
	1	8	2	4	5	8

Binary Clock: *More of a gimmick than a useful device, this binary clock expresses the sexagesimal time by representing each of the decimal digits for hours, minutes, and seconds in binary code. This system is known as 'binary code decimal', or BCD. A true binary clock would display the values for hours, minutes, and seconds in just three single columns that include 16s and 32s.*

Measuring Firewood

Throughout North America, firewood is sold by the 'cord'. In most states and provinces, this is a legally defined quantity. In Canada, for example, a cord is defined by the Weights and Measures Regulations as '128 cubic feet of stacked roundwood (whole or split, with or without bark) containing wood and airspace with all bolts of similar length piled in a regular manner with their longitudinal axes approximately parallel.'

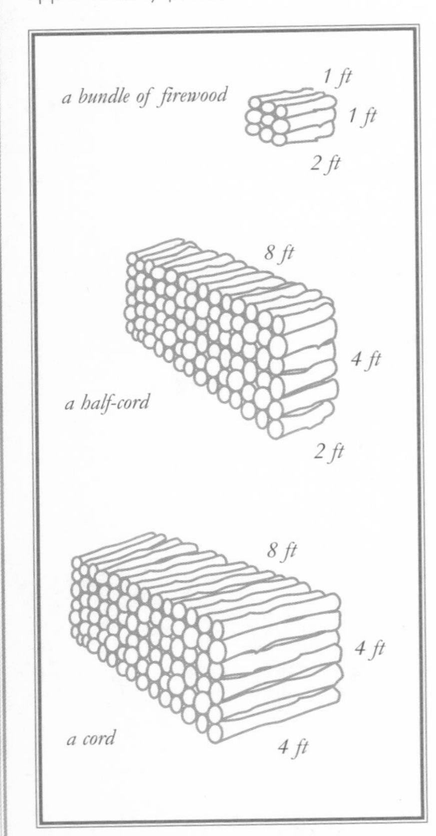

In metric terms, a cord is therefore equal to 3.62454 cubic metres. Buyers are advised to stack the wood (ideally in an easily assessed block measuring 4 ft × 4 ft × 8 ft) to check whether they have received a full measure. A cord foot is one-eighth of this, the volume of a stack of firewood 4 ft × 4 ft × 1 ft (16 cubic feet, or 0.4531 cubic metres).

In addition, buyers are warned to avoid buying wood in quantities such as a face cord, rack, rick, tier, pile, or truckload, as these terms are not legally defined units of measure. Some of them do, nonetheless, have traditional definitions. The face cord, for example, is usually one-third of a cord, or a stack of 16-inch logs 4 ft high and 8 ft long, although in some parts of America the face cord is half a cord. A rick, derived from the Old Norse meaning an organised stack, is the same as a face-cord. A bundle is a very vague term, but usually means about two cubic feet, or 1/64 of a cord.

European Firewood

The European unit *stère* – which is derived from the Greek word *stereos*, meaning 'solid' – has been a unit of volume in the metre–tonne–second or mts system of units since the French Revolution. However, its use is now limited to a unit of firewood. One stère (st) is equal to one cubic metre (or 0.2795 cords).

How Hot Is a Chilli?

The hot sensation that comes from eating a chilli is caused by a chemical called capsaicin. Found in all the members of the chilli family, this acts on sensory neurons called vanilloid receptors, which also respond to high temperatures. Capsaicin produces such a strong effect that it will desensitise pain receptors, and can even kill them. This pain-killing effect has led to the use of the chemical in ointments to reduce muscle pain that produce a 'warm' sensation in the skin.

Scoville Rating

So how can the hotness of chilli be measured? In 1912, chemist Wilbur L. Scoville (1865–1942), working for Parke Davis pharmaceuticals, came up with the Scoville Organileptic Test to measure the 'hotness' of a chilli. He did this by having a panel of tasters (usually five people) determine whether the presence of the chilli flavour was detectable in ever more dilute solutions. The Scoville rating is given by the dilution at which it can no longer be tasted. Sweet peppers, which contain no capsaicin, have a Scoville rating of zero, and the scale goes up through Tabasco Pepper Sauce (2,500–5,000), jalapeño peppers (2,500–8,000), serrano peppers (10,000–23,000), scotch bonnets (100,000–325,000), and red savina habaneros (350,000–577,000). Pure capsaicin has a Scoville rating of 16,000,000, which means you could taste the presence of half a cupful in an Olympic swimming pool.

Modern Methods

The downside of the Scoville system is that it is highly subjective – individuals have differing sensitivity, and continued tasting can lead to the receptors experiencing a build-up of heat or becoming desensitised. Using techniques such as high-pressure liquid chromatography (HPLC), the concentration of capsaicin can be measured directly, giving much more accurate results.

Habanero: *One of the hottest chilli peppers in the world, measuring up to 577,000 on the Scoville scale.*

Pest Control

The chilli family probably evolved capsaicin because it offers protection against being eaten by herbivores. Birds are largely insensitive to its presence, so it can be put in bird feeders to discourage squirrels. Capsaicin is also used in products to prevent horses from chewing their rugs, and pepper spray is commonly carried by hikers in grizzly-bear country. A similar spray is used in some parts of the world to disperse riots. The chemical name for capsaicin is 8-methyl-N-vanillyl-6-nonenamide, and its molecular formula is $C_{18}H_{27}NO_3$.

Ring Sizes

As anyone who has ever tried to buy a ring for a loved one while travelling abroad may have discovered, not every country uses the same system of ring sizes. In fact, the diversity of systems is an eloquent testimony to human ingenuity in the field of measurement.

The only factor that determines whether a ring will fit the wearer's finger is the size of the hole in the middle. Assuming that this is circular, there are two ways of defining this: the internal circumference, and the internal diameter. Since the diameter gives a smaller measurement, circumference is the more accurate, and this measurement – in millimetres – is the basis of the European metric system. However, there are at least another four 'standard' ways to size a ring.

American Sizes

North American ring sizes are based on the numbers 0 to 16, interspaced with intermediary half and quarter sizes. The difference between two successive whole integer sizes was previously 1/10 of an inch of internal circumference, but since the standardisation of the U.S. and Imperial inches by reference to the metre, the difference is now 2.6 mm of internal circumference.

Other Sizes

Developed in the early 1900s by Joseph Pepper, an English jeweller, the 'Wheatsheaf' scale works on a system running from A to Z, with evenly spaced intermediate half sizes. In 1945, the British Standard was based on the Wheatsheaf gauge, with divisions between sizes in steps of 1/64 of an inch of internal diameter. In 1987, a new standard was introduced based on SI units, with one (alphabetic) size division equal to 1.25 mm of internal circumference. Britain, Ireland, and Australia all use this system. One whole size U.K. is roughly equivalent to one half-size U.S.

The Japanese system is based on internal ring diameter, size one being 13 mm, and each division being one-third of a millimetre.

Some European countries use ring sizes given by the internal circumference in mm minus 40, making a 44 mm ring a size 4.

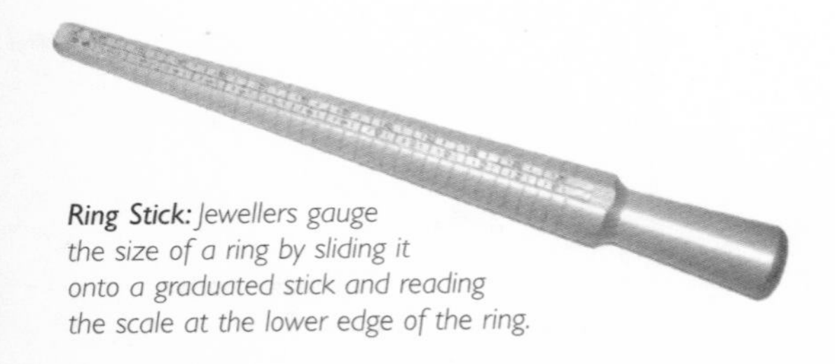

Ring Stick: *Jewellers gauge the size of a ring by sliding it onto a graduated stick and reading the scale at the lower edge of the ring.*

Finger Sizes (right): *A set of numbered blank rings is used by jewellers to find out what size of ring is required, ensuring the ring goes over the knuckle.*

Ring-Size Chart

The first two columns give the size by diameter; the next two by circumference. The next four columns show the sizes used in the U.S./Canada (Z1), the Wheatsheaf sizes used in the U.K., Ireland, and Australia (Z2), and then the Japanese (Z3) and European (Z4) sizes.

D(in)	D(mm)	C(in)	C(mm)	Z1	Z2	Z3	Z4	D(in)	D(mm)	C(in)	C(mm)	Z1	Z2	Z3	Z4
0.458	11.6	1.438	36.5	0				0.722	18.3	2.267	57.6	8 1/4	Q		
0.466	11.8	1.463	37.2	1/4				0.730	18.5	2.292	58.2	8 1/2	Q 1/2	17	
0.474	12.0	1.488	37.8	1/2	A			0.738	18.7	2.317	58.9	8 3/4	R		19
0.482	12.2	1.513	38.4	3/4	A 1/2			0.746	18.9	2.342	59.5	9	R 1/2	18	
0.490	12.4	1.539	39.1	1	B	1		0.754	19.2	2.368	60.1	9 1/4	S		20
0.498	12.6	1.564	39.7	1 1/4	B 1/2			0.762	19.4	2.393	60.8	9 1/2	S 1/2	19	
0.506	12.9	1.589	40.4	1 1/2	C			0.770	19.6	2.418	61.4	9 3/4	T		
0.514	13.1	1.614	41.0	1 3/4	C 1/2		1	0.778	19.8	2.443	62.1	10	T 1/2	20	22
0.522	13.3	1.639	41.6	2	D	2		0.786	20.0	2.468	62.7	10 1/4	U	21	
0.530	13.5	1.664	42.3	2 1/4	D 1/2			0.794	20.2	2.493	63.3	10 1/2	U 1/2	22	23
0.538	13.7	1.689	42.9	2 1/2	E	3	3	0.802	20.4	2.518	64.0	10 3/4	V		
0.546	13.9	1.714	43.5	2 3/4	E 1/2			0.810	20.6	2.543	64.6	11	V 1/2	23	
0.554	14.1	1.740	44.2	3	F	4	4	0.818	20.8	2.569	65.2	11 1/4	W		25
0.562	14.3	1.765	44.8	3 1/4	F 1/2	5	5	0.826	21.0	2.594	65.9	11 1/2	W 1/2	24	
0.570	14.5	1.790	45.5	3 1/2	G			0.834	21.2	2.619	66.5	11 3/4	X		
0.578	14.7	1.815	46.1	3 3/4	G 1/2	6	6	0.842	21.4	2.644	67.2	12	X 1/2	25	27
0.586	14.9	1.840	46.7	4	H	7	7	0.850	21.6	2.669	67.8	12 1/4	Y		
0.594	15.1	1.865	47.4	4 1/4	H 1/2			0.858	21.8	2.694	68.4	12 1/2	Z	26	
0.602	15.3	1.890	48.0	4 1/2	I	8		0.866	22.0	2.719	69.1	12 3/4	Z 1/2		29
0.610	15.5	1.915	48.7	4 3/4	J		9	0.874	22.2	2.744	69.7	13		27	
0.618	15.7	1.941	49.3	5	J 1/2	9		0.882	22.4	2.769	70.3	13 1/4	Z1		
0.626	15.9	1.966	49.9	5 1/4	K		10	0.890	22.6	2.795	71.0	13 1/2			
0.634	16.1	1.991	50.6	5 1/2	K 1/2	10		0.898	22.8	2.820	71.6	13 3/4	Z2		
0.642	16.3	2.016	51.2	5 3/4	L		11	0.906	23.0	2.845	72.3	14	Z3		
0.650	16.5	2.041	51.8	6	L 1/2	11	12	0.914	23.2	2.870	72.9	14 1/4			
0.658	16.7	2.066	52.5	6 1/4	M	12		0.922	23.4	2.895	73.5	14 1/2	Z4		
0.666	16.9	2.091	53.1	6 1/2	M 1/2	13		0.930	23.6	2.920	74.2	14 3/4			
0.674	17.1	2.116	53.8	6 3/4	N		14	0.938	23.8	2.945	74.8	15			
0.682	17.3	2.141	54.4	7	N 1/2	14		0.946	24.0	2.970	75.4	15 1/4			
0.690	17.5	2.167	55.0	7 1/4	O		15	0.954	24.2	2.996	76.1	15 1/2			
0.698	17.7	2.192	55.7	7 1/2	O 1/2	15	16	0.962	24.4	3.021	76.7	15 3/4			
0.706	17.9	2.217	56.3	7 3/4	P			0.970	24.6	3.046	77.4	16			
0.714	18.1	2.242	56.9	8	P 1/2	16	17								

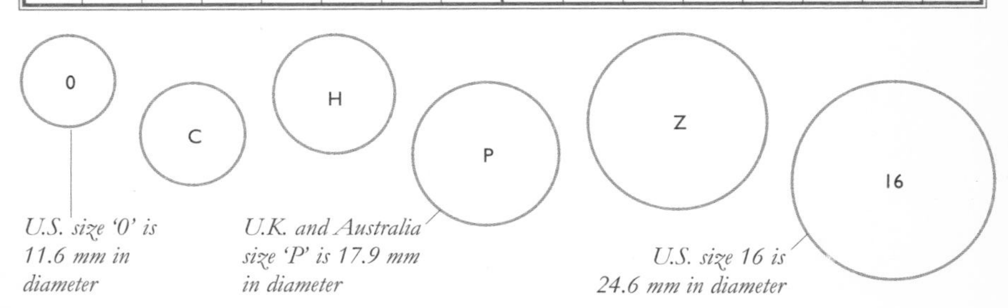

U.S. size '0' is 11.6 mm in diameter

U.K. and Australia size 'P' is 17.9 mm in diameter

U.S. size 16 is 24.6 mm in diameter

Units of Concentration

In science, healthcare, manufacturing, or everyday life, there is a constant need to know what proportion of a given substance is present in a solid, liquid, or gas, whether it be a medical drug or an alcoholic drink, a blood sample or polluted air.

Chemical Concentrations

The need for accuracy in chemical compounds and mixtures demands precise means for expressing the concentrations of solutions. The dissolved material is called the 'solute', and the medium in which it is dissolved is called the 'solvent'. A solute may be a soluble solid, such as table salt, or it may be another liquid.

There are several ways of describing the concentration of the resulting mixture. Several of these are based on the SI unit for amount of substance: the mole.

Molarity (M): Molarity is the number of moles of solute per litre of solution. (Since the addition of the solute can affect the total volume, 'per litre of solution' is not the same as 'per litre of solvent'.) If a litre of solution contains 0.5 moles of solute, the solution has a molarity of 0.5.

Molality (m): This is the number of moles of solute per kilogram of solvent. If the solvent is water at about 77 °F (25 °C), and the solution is weak (the solute adding little to the density of the solution), then a litre of solution will weigh almost exactly 1 kg, in which case the molality and molarity will be the same. For a concentrated solution, or one in which the solvent is more or less dense, the two are different.

Mole Fraction (X): The ingredients in any solution can be expressed as the number of moles of each. The mole fraction of any solute is the number of moles of that chemical expressed as a fraction of the total number of moles of all chemicals in the solution. For example, if a solution consisted of one mole of sodium chloride and nine moles of water, the mole fraction of sodium chloride would be 1 mol/1 + 9 mol = $^1/_{10}$ = 0.1.

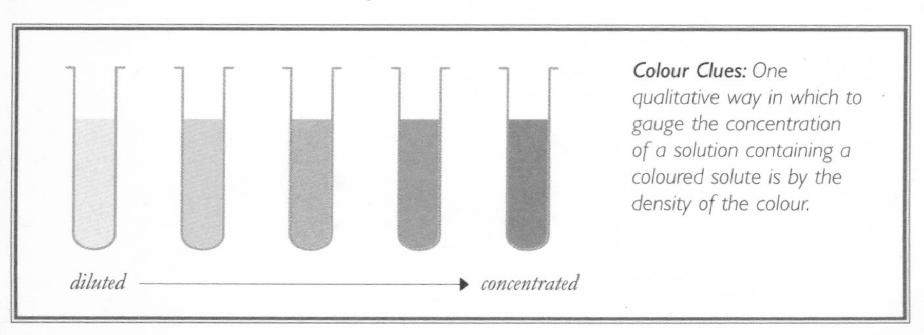

Colour Clues: *One qualitative way in which to gauge the concentration of a solution containing a coloured solute is by the density of the colour.*

Alcohol Content

In Britain in the seventeenth and eighteenth centuries, prior to the introduction of the hydrometer to test the alcohol (specifically ethanol) content of liquor, the 'proof' method was used. A little of the liquid was added to gunpowder – if the powder could still be ignited, the liquor was considered 100 proof. It was later found that 100 proof was equivalent to 57.15% ethanol, making 70 proof equal to 40% alcohol by volume (abv or alc/vol). This definition of proof is still in use in the U.K., although, by agreement with the rest of the European Community, the alcohol by volume must be stated on alcoholic drinks. Proof is optional.

In the U.S., too, the abv must always be indicated, but proof is sometimes also given. The U.S. definition of proof is twice the alcohol by volume, so 70 proof in the U.S. is 35% abv.

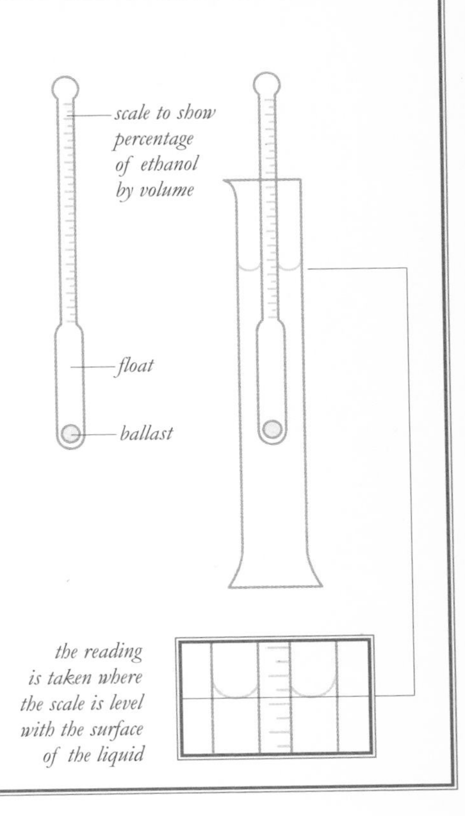

AMOUNT PER VOLUME

In medical contexts, the concentration of specific substances is often expressed as the amount of the substance (by mass or moles) per unit volume. For example, people with diabetes regularly test their blood in order to keep a close watch on their blood sugar levels. This is a measure of the concentration of glucose in the blood, also known as the serum glucose level. In the U.S. it is expressed as milligrams per decilitre (mg/dL), while in the U.K. and Canada this figure is given as millimoles per litre (mmol/L). The normal range is approximately 72 to 144 mg/dL (4 to 8 mmol/L), but levels tend to be lower in the morning and higher after meals.

PERCENT COMPOSITION BY MASS (%)

As its name suggests, this is an expression of the mass of the solute divided by the total mass of the solution, so if 1 kg of solution contains 10 g of copper sulphate, this is a one percent copper sulphate solution.

Hardness

Hardness can mean several different things, such as the resistance to being scratched, dented, abraded, drilled, or cut. The kind of hardness being measured will depend on the method of measurement, but they are all measures of force withstood.

Friedrich Mohs
1773–1839

Scratch Tests

Early in the nineteenth century, the German mineralogist Friedrich Mohs devised a measure of mineral hardness based on the 'scratchability' of ten well-known minerals. Although it is referred to as Mohs's Scale, it is in fact a ranking that compares the hardness of various minerals, rather than a scale that gives a quantitative value. This is because the interval between numbers is not constant. Corundum (for example, sapphire and ruby) is twice as hard as topaz, while diamond is four times as hard as corundum.

Mohs's Scale

Mineral	Hardness
Diamond	10
Corundum	9
Topaz	8
Quartz	7
Feldspar	6
Apatite	5
Fluorite	4
Calcite	3
Gypsum	2
Talc	1

It is nonetheless a useful identification tool for mineralogists in the field. If an unknown specimen scratches a known mineral from the list, it is harder than that mineral. If it is scratched by another known mineral, it is softer than that mineral.

Scratchability is measured more quantitatively by Turner's Sclerometer, in which a weighted diamond point is drawn across the smooth surface of the material to be tested. The hardness number is the weight in grams required to produce a 'standard' scratch.

Dent Tests

Hardness can be measured by a material's resistance to being crushed. Brinell's Test, devised by J. A. Brinell in 1900, is conducted using an apparatus that applies a force of 6,600 lb (3,000 kg) to a 0.4 inch (10 mm) hardened steel ball to press it into the smooth surface of the material being tested. From the area of the spherical indentation, a hardness value is calculated. This method is commonly used to test metals, but it requires the test sample to be placed on the bed of the testing apparatus.

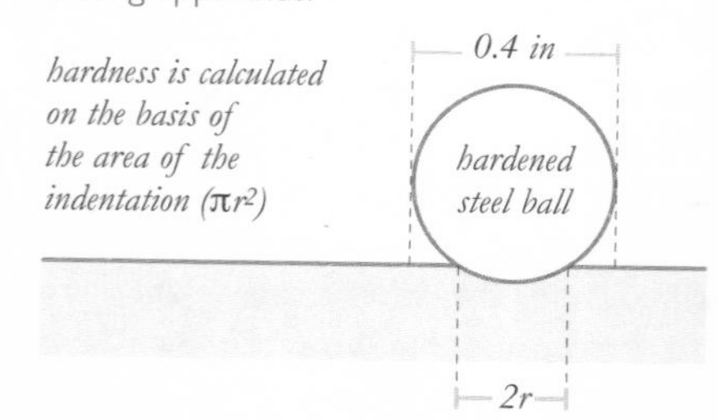

Microhardness

The Knoop method uses a rhombic-based pyramidal diamond indenter that forms a very small elongated indent. The sample material has to be finely polished for this method to work, and a microscope is used to determine the maximum length of the indent, which is used to determine the Knoop hardness.

Due to the small size of the indent, this method is particularly useful for measuring the hardness of a thin layer or of brittle material.

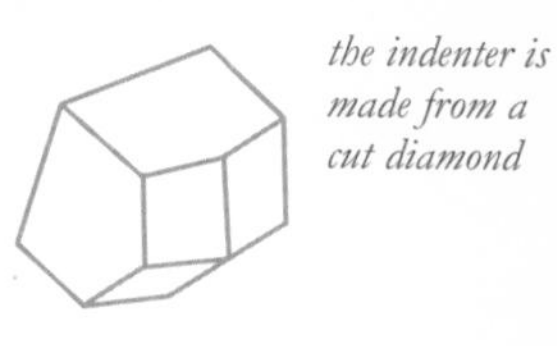

the indenter is made from a cut diamond

length of indent determines hardness

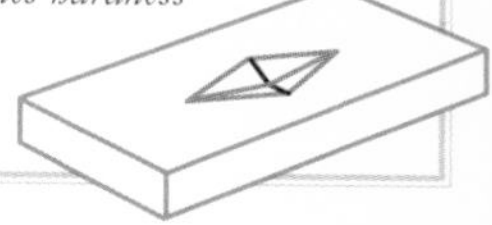

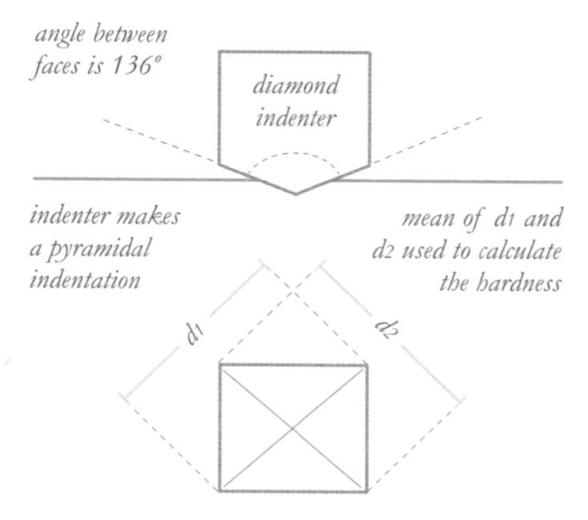

The Vickers Hardness Test method works in a similar way, using a diamond indenter that has a pyramidal point with an angle of 136° between opposite faces. The area of the sloping surface of the indentation is divided by the applied force to give the Vickers hardness number of the material.

The Rockwell Hardness Test is similar, but uses an initial load, followed by a greater load, and then a return to the smaller initial load. The extent of the final indentation is then measured. This is the most commonly used method of testing.

Dynamic Hardness

Another measure of the hardness of a substance is the degree to which an object will bounce off it. Shore's Scleroscope, introduced in the early 1900s, measures this quantitatively by dropping a small pointed steel cylinder (known as the 'tup' or 'tuppet') onto the smooth surface of the material to be tested, and measuring the height of the rebound. One of the advantages of this measuring device is that it is portable and can therefore be used on materials in situ, rather than having to bring them into a laboratory.

Drilling Through

Attempting to drill a hole in a material gives an immediate indication of its hardness. Keep's Test involves pressing a standard steel drill with a standard force against the material and turning the drill a specified number of turns. The angle of a line representing the rate at which the drill penetrates the material gives a measure of the hardness.

Typography

'Point' is a unit of measure used in typography to express the size of a font. There are 12 points to the pica, and 72 points to the inch (making 6 picas to the inch). This system of measurement arose in the era of movable lead type, when typesetters ceased making their own type and there was a need to standardise the individual lead letters being made by different manufacturing companies. At the time, each individual letter was a small block of lead with the letter embossed on one face. For the blocks to fit together in a single line, these blocks needed to be the same size, and so it was necessary to define the 'body size' – the size of the lead block – for each font. The block was always large enough to bear an ascending or descending letter, with a slight gap at top or bottom.

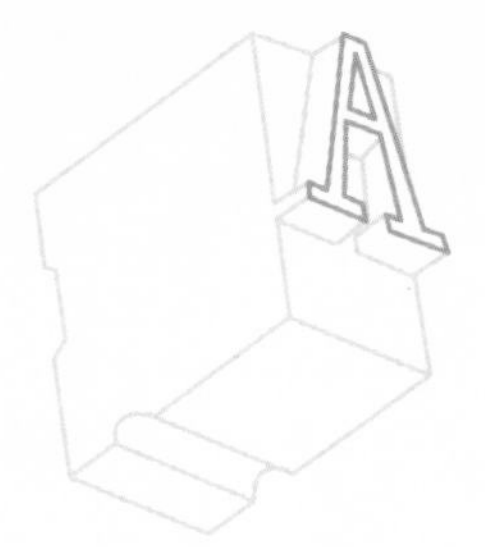

type block with embossed letter

the body size of a font was defined by the size of the block on which it was set

Typefounders

French clergyman Sébastien Truchet (1657–1729) came up with a unit that was the lowest common denominator of all the current sizes – 1/1,728 of a pied du Roi (Royal foot) – but it did not become established. The type founder Pierre Simon Fournier (1712–1768) later used a typographical point of about 1/864 of a royal inch, but it was not until François-Ambroise Didot (1730–1801) based his point on that of Truchet but doubled its size to make the point 11/864 of a royal foot (1/72 of a royal inch), that the system found widespread popularity. Indeed, Didot's system became the standard throughout Continental Europe. In metric units, the point Didot is equal to 0.3759 mm, and 12 points equals one Cicero.

In the U.S., the American point was defined in 1886 by the Type Founders' Association of the United States and the Monotype Corporation as 1/72.272 inch, just slightly different from the Continental Didot point. This was taken up in Britain by the end of the nineteenth century, at which point it became known as the Anglo-American point.

Digital Typography

By the beginning of the 1980s, lead type had virtually disappeared from use, as photocomposition and computers began to take over. The new Tex electronic typesetting language used a point size of 1/72.27 inch that was almost identical to the Anglo-American and Monotype points, but with the advent of the Apple Macintosh computer and a monitor with exactly seventy-two pixels per inch, the computer point was made exactly 1/72 inch. This point size – also known as the Postscript point – is now used in all forms of digital typography.

Defining Type Size

It can be argued that the point is not needed as a unit of measurement. After all, there is no reason why type size and spacing cannot be expressed in SI units such as the millimetre, in the same way as all other aspects of page layout. Indeed, in 1979, the 16th General Conference on Weights and Measures stated that the proliferation of special names represents a danger for SI and 'must be avoided in every possible way.' One perceived downside of the way in which the size of type is currently defined is that the measurement relates to an invisible grid – no aspect of the printed type corresponds to its actual size, not even the distance between the top of an ascender and the bottom of a descender. As a result, there is a movement to redefine the basis on which type size is calculated and find a definition that truly reflects the visual impression that a font makes. In 1975, a working group of the International Organisation for Standardisation suggested using the height of the capital letter as the basis of type size, but this came to nothing. The x-height – which is arguably the most important single factor determining the look of type on the page – is another alternative, but the debate continues.

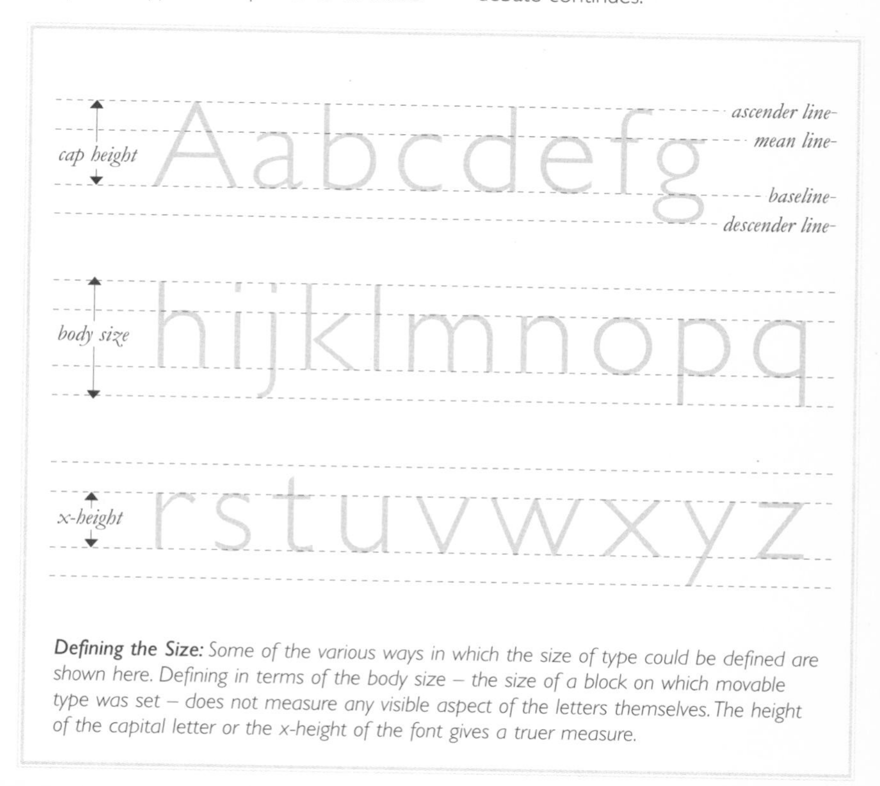

Defining the Size: *Some of the various ways in which the size of type could be defined are shown here. Defining in terms of the body size – the size of a block on which movable type was set – does not measure any visible aspect of the letters themselves. The height of the capital letter or the x-height of the font gives a truer measure.*

SI Diagram

If you think the page opposite looks like an incomprehensible dry-as-dust diagram, then look again. This is a map of the all the relationships between the units of the SI system, but is much more than just that.

The State of Play

This is a diagrammatic statement of our present understanding of the way the universe works, of the relationships that have been observed to exist between length and mass, time, electrical current, temperature, matter, and light. We are indebted for this cerebral work of art to the U.S. National Institute of Standards and Technology, and I would advise anyone to take a look at the NIST website (http://physics.nist.gov/). I would also recommend a visit to the sites of the U.K. National Physical Laboratory (www.npl.co.uk/reference/) and the International Bureau of Weights and Measures (www.bipm.fr/en/home/), as all of these offer valuable insights and information on the SI system of units.

Reading the Map

The seven base units of SI are shown as rectangles in the left-hand column. In each case, the name, symbol, and base quantity (such as length) are shown. In the second column are the units derived directly from the base units, shown as double circles. These do not have special names other than descriptive terms (such as metre squared). Some of these are derived from more than one base unit (such as metre per second). In this case a solid arrow brings in the base unit that is the nominator (metre) and a dotted arrow brings in the denominator (per second). The third column contains the derived units with special names, shown as circles, and again a solid arrow indicates multiplication and a dotted arrow indicates division.

Clear and Simple

These seven base units and 22 derived units are the full complement of the SI system, demonstrating perfectly its advantages. Taking energy as an example, there is only one unit – the joule – whereas in the imperial and U.S. Customary systems energy is measured in various different units, including British Thermal Units, calories, kilowatt hours, and kilotons, which obscures the fact that chemical, electrical, mechanical, and thermal energy are all equivalent.

Unconnected Units

There are two units – the radian and the steradian – that have no arrows connecting them to any SI base units. This is because plane angle is expressed as the ratio of two lengths (radius and arc), and solid angle as the ratio of an area and the square of a length (the radius squared). The derived units are therefore m/m and m^2/m^2, making the coherent derived unit 1 in both cases. They are therefore dimensionless. The special names radian and steradian are given for clarity.

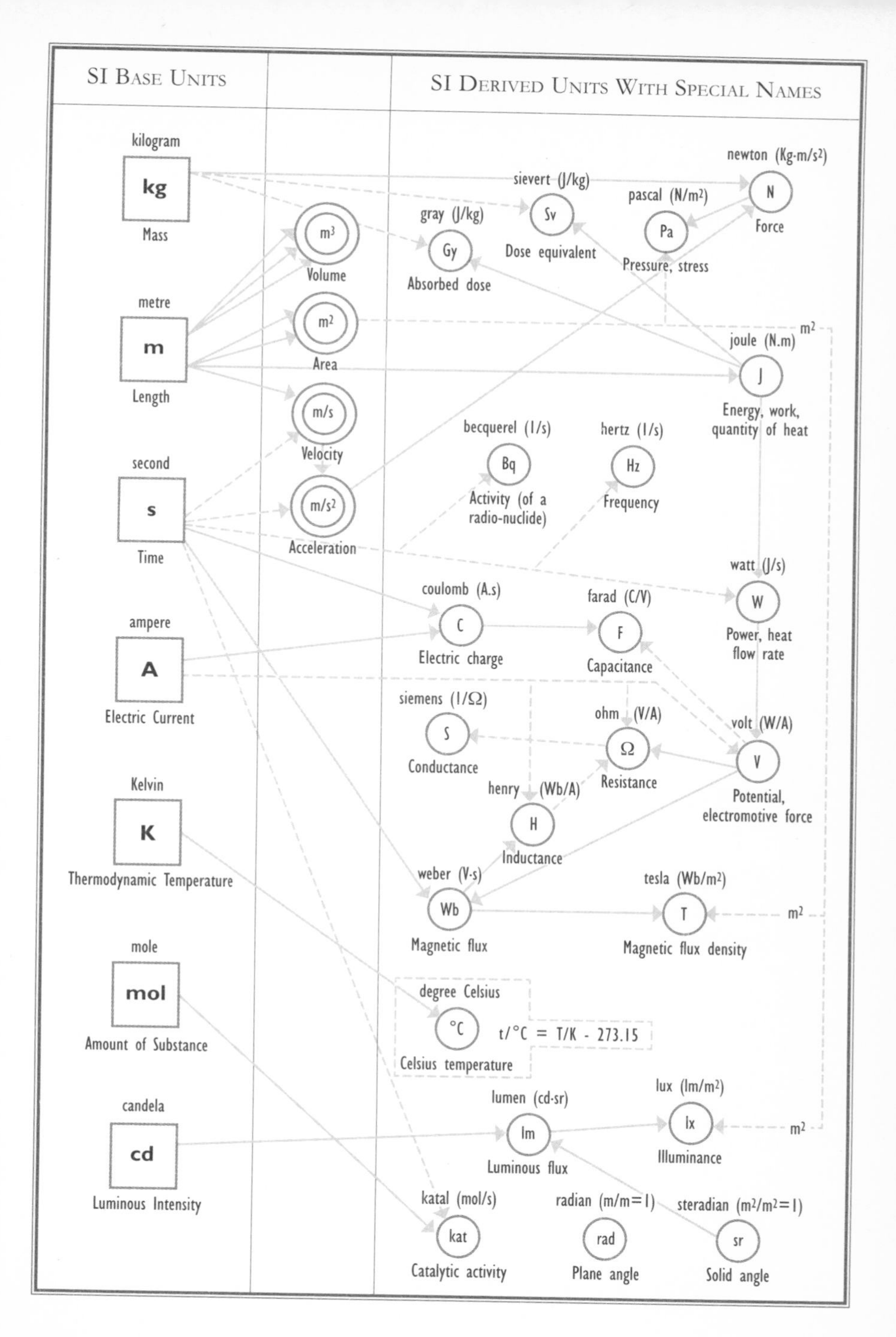

SI Base Units
SI Derived Units With Special Names
kilogram
kg
Mass
metre
m
Length
second
s
Time
ampere
A
Electric Current
Kelvin
K
Thermodynamic Temperature
mole
mol
Amount of Substance
candela
cd
Luminous Intensity
m³
Volume
m²
Area
m/s
Velocity
m/s²
Acceleration
newton (Kg·m/s²)
N
Force
sievert (J/kg)
Sv
Dose equivalent
pascal (N/m²)
Pa
Pressure, stress
gray (J/kg)
Gy
Absorbed dose
m²
joule (N.m)
J
Energy, work, quantity of heat
becquerel (1/s)
Bq
Activity (of a radio-nuclide)
hertz (1/s)
Hz
Frequency
watt (J/s)
W
Power, heat flow rate
coulomb (A.s)
C
Electric charge
farad (C/V)
F
Capacitance
siemens (1/Ω)
S
Conductance
ohm (V/A)
Ω
Resistance
volt (W/A)
V
Potential, electromotive force
henry (Wb/A)
H
Inductance
weber (V·s)
Wb
Magnetic flux
tesla (Wb/m²)
T
Magnetic flux density
m²
degree Celsius
°C
Celsius temperature
t/°C = T/K - 273.15
lumen (cd·sr)
lm
Luminous flux
lux (lm/m²)
lx
Illuminance
m²
katal (mol/s)
kat
Catalytic activity
radian (m/m=1)
rad
Plane angle
steradian (m²/m²=1)
sr
Solid angle

Epilogue

This has been a whistle-stop tour through time and space, touching on some very diverse aspects of our everyday lives as well as a host of scientific topics. Are there any conclusions that we can draw from the journey?

Touching on Fundamentals

If there is one thing that this book should have brought home it is that units of measurement are neither arbitrary nor irrelevant. At any point in human history, the units in use are a reflection of the way the people who use them see and understand the world. They are not merely a means of recording the details of the objects and phenomena around us; they are an integral part of knowledge itself. The complexity and coherence of modern units of measurement are themselves a measure of our understanding of the way the universe works.

The Tip of the Iceberg

Throughout this book we have tried to be as comprehensive as possible, to cover the broad history of the development of units and means of measurement, to highlight some of the individuals whose scientific work has deepened our understanding of the world around us, and to explain the origins, workings, meanings, and relationships of the principal units and systems in use today. We hope that we have thrown some light on these topics, but we're aware that we have only scratched the surface.

For every subject we have covered, there are experts – on the history, on the details, on the mathematics, and especially on the physics – whose individual knowledge would fill this book a dozen times over. To move to the next level of discussion and explanation would necessitate the inclusion of reams of academic debate, of complex formulae and algebraic equations, and would require the reader to have a working understanding of every branch of science, from astronomy and electromagnetism to spectroscopy, optics, and quantum physics.

Taking It Further

If this book has piqued your interest and you want to know more, the information is out there – in text books, in academic journals and, in abundance, on the Internet. Just key in any measurement-related keyword and you will be met by a barrage of sites offering definitions, histories, discussion forums, political agendas, fierce arguments, and a wealth of erudite scientific explanation. In addition, there are countless businesses and industries manufacturing and selling every conceivable form of measuring equipment, as well as university departments, research institutions, and a host of national and international organisations whose purpose is to monitor, regulate, coordinate, promote, research, and improve systems of measurement. Many websites press for the complete worldwide adoption of SI, and the naysayers seem bound to lose the battle.

The Science of Metrology

Measurement is now a science in its own right, and this science is called 'metrology'. In the words of the Director of the BIPM, Professor A. J. Wallard, metrology 'embraces both experimental and theoretical measurements and determinations at any level of uncertainty in any field of science and technology. The intricate and mostly invisible networks of services, suppliers and communications upon which we are all dependent rely on metrology for their efficient and reliable operation.'

Every year, the 51 Members of the Metre Convention and the 17 Associate States and Economies of the General Conference on Weights and Measures celebrate Metrology Day on May 20th, the date on which the Convention was signed in 1875. In 2005, on the Convention's 130th birthday, Professor Wallace took the opportunity to highlight the important role that metrology plays in so many aspects of our lives.

A Central Role

In commerce, manufacturing, and especially engineering, accurate and reliable measurement is central to quality assurance, economic success, and safety. Without the accurate measuring of time and its international coordination, satellite navigation and global computer networking would not be possible. The health service relies on measurement for diagnosis and treatment, especially in fields such as radiotherapy and laboratory medicine. Meteorology and global environmental monitoring have measurement at their cores. These are just a few examples. As for the future, he sees the challenges coming from physics and engineering, but especially from the chemistry-based sciences.

It seems appropriate to close this book with his concluding words: 'The adventure of metrology is an enterprise which has been propelling the evolution of the modern world and which continues to excite the imagination and to assist society.'

15 16 17 18 19 20 21 22 23 24 25 26 27 28

6 7 8 9 10 11

Index